LEAN Production Training for Practice and Study—35 Exercises with Solutions

AF809089

Roman Hänggi · Patrick Balve · Lukas Budde

# LEAN Production Training for Practice and Study—35 Exercises with Solutions

Based on the Book "LEAN Production—Easy and Comprehensive"

Springer Vieweg

Roman Hänggi
Brülisau, Inner-Rhoden, Switzerland

Patrick Balve
Leingarten, Germany

Lukas Budde
Abtwil, Switzerland

ISBN 978-3-662-70467-7 ISBN 978-3-662-70468-4 (eBook)
https://doi.org/10.1007/978-3-662-70468-4

Translation from the German language edition: "LEAN Production Training für Praxis und Studium – 35 Übungen mit Lösungen" by Roman Hänggi et al., © Der/die Herausgeber bzw. der/die Autor(en), exklusiv lizenziert an Springer-Verlag GmbH, DE, ein Teil von Springer Nature 2024. Published by Springer Berlin Heidelberg. All Rights Reserved.

© The Editor(s) (if applicable) and The Author(s), under exclusive license to Springer-Verlag GmbH, DE, part of Springer Nature 2025

This work is subject to copyright. All rights are solely and exclusively licensed by the Publisher, whether the whole or part of the material is concerned, specifically the rights of translation, reprinting, reuse of illustrations, recitation, broadcasting, reproduction on microfilms or in any other physical way, and transmission or information storage and retrieval, electronic adaptation, computer software, or by similar or dissimilar methodology now known or hereafter developed.
The use of general descriptive names, registered names, trademarks, service marks, etc. in this publication does not imply, even in the absence of a specific statement, that such names are exempt from the relevant protective laws and regulations and therefore free for general use.
The publisher, the authors and the editors are safe to assume that the advice and information in this book are believed to be true and accurate at the date of publication. Neither the publisher nor the authors or the editors give a warranty, expressed or implied, with respect to the material contained herein or for any errors or omissions that may have been made. The publisher remains neutral with regard to jurisdictional claims in published maps and institutional affiliations.

This Springer Vieweg imprint is published by the registered company Springer-Verlag GmbH, DE, part of Springer Nature.
The registered company address is: Heidelberger Platz 3, 14197 Berlin, Germany

If disposing of this product, please recycle the paper.

# Preface

This book would not have been possible without the fundamental textbook *LEAN Production—Easy and Comprehensive* by Roman Hänggi, André Fimpel, and Roland Siegenthaler. The presentation of Lean production in a book with many practical examples and tips has been very well received by companies and students. The presentation of Lean theory and its implementation in practice is simple and comprehensive. The Lean concept is also explained clearly, precisely, and quickly in over 200 pictures.

The idea for this in-depth exercise book stems from Patrick's contact with Roman, during which Patrick praised the book *LEAN Production—Simple and Comprehensive* as a very understandable and comprehensive textbook for teaching. In his opinion, it is the presentation of Lean principles and methods using a consistent company example, LeanClean Inc., that makes the book so unique and valuable for his students at Heilbronn University. In addition, Lukas, the third author in the group, uses the book as a source of ideas for exercises in the "Production Management" lecture at the University of St. Gallen. The practical relevance is a great help for the students. Roman has developed a wide range of Lean exercises based on the book for his classes at the University of Applied Sciences of Eastern Switzerland OST. Thanks to the students' feedback, he improved the exercises from year to year.

The discussion among the three authors about Lean education in university teaching revealed that the concrete "doing" of Lean in the classroom is central to the learning effect. The executive education programs at the University of St. Gallen and various practical projects have also shown Roman that the exercises are also important for training and understanding Lean in practice and especially in management.

The numerous responses from practitioners and students were motivation enough to make these exercises available to a wider circle of people interested in Lean: The idea for this practical exercise book was born.

Theory thrives on implementation. This includes practicing or directly applying the theory. Nevertheless, the book begins with an introductory chapter on theory. The purpose of this first part is not to present Lean theory comprehensively, but to introduce the basic mindset—the elimination of waste in its various forms. The main part of the book consists of 35 exercises. They deepen the theory, especially the application of the relevant methods, of the textbook *LEAN Production—Simple and Comprehensive*. Each exercise consists of a number of tasks and a sample solution, which makes it easy to study the content on your own. Wherever we felt it was appropriate, we have repeated selected content from the textbook in the exercises and added smaller theoretical passages and practical tips.

We now hope that many students at universities and colleges can deepen their understanding of the concepts and methods of Lean production with these tasks and successfully implement them after their graduation. Further application examples are also very welcome in companies to deepen the understanding of Lean and to train the implementation. The independent "doing" based on these exercises can thus become the starting point for a successful implementation of Lean in factories and administration.

We hope you enjoy applying Lean theory. In the spirit of "there's nothing good unless you do it"—let's get started on "doing Lean".

Brülisau, Switzerland
Leingarten, Germany
Abtwil, Switzerland

Roman Hänggi
Patrick Balve
Lukas Budde

# Introduction

The book *LEAN Production—Simple and Comprehensive* is popular in both in practice and in teaching, as it explains the complexity and variety of lean topics through a consistent, practical example. This comprehensive presentation helps readers understand why and how Lean can unlock significant potential for operational excellence, as many companies have experienced in recent years. These companies have also learned that the Lean journey never ends—there are always further improvements to be made and opportunities to become an even better company.

Over the past three decades, numerous Lean implementers have learned that Lean production is not limited to the automotive industry. Lean concepts can be applied across all types of industrial production and are increasingly relevant in other sectors, such as construction and services. The details of each Lean program are, of course, company and industry specific. While the specifics of each Lean program are unique to the company and industry, the central elements remain the same: principles for analyzing waste, principles for designing low-waste processes, and concrete methods for implementing both. The structure of this exercise book also follows the same logic as the underlying textbook.

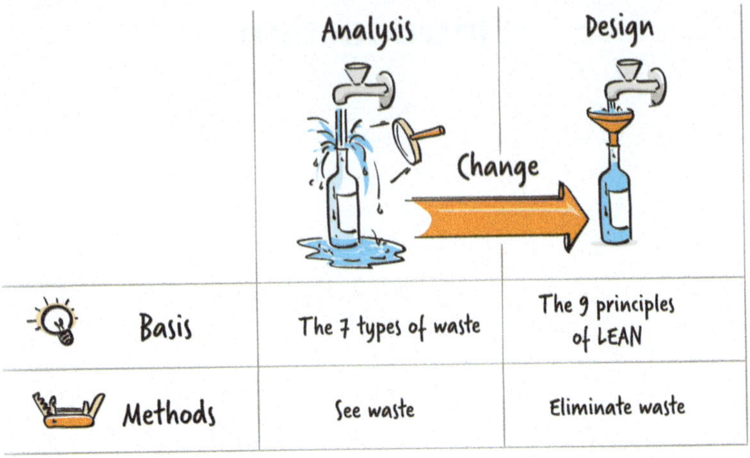


The Lean journey starts with understanding the core principles of Lean, including the seven types of waste and the nine principles of Lean, which form a comprehensive foundation for applying Lean methods. Understanding these basics is important before implementing Lean practices within a company, as this understanding is necessary at all hierarchical levels. Once these principles are grasped, relevant Lean methods can be applied to drive the required changes within your company. A targeted, sequential approach is important for effective implementation, as this leads to change management. Lean transformation fundamentally changes the corporate culture, which is crucial for successfully embedding Lean practices over the long term.

The first challenge in implementing Lean effectively is the lack of comprehensive discussion and understanding of its principles. Lean is often perceived as a "logical" concept, leading people to underestimate the importance of taking the time to train and engage in discussions about it. However, thorough exploring of Lean's content, implications, and the changes required for a specific company is crucial. Through this process, it often becomes apparent that Lean is not as simple as it seems; it raises many questions and is more challenging to implement than initially expected. This change is only possible if top management understands Lean in all its facets and if this knowledge is broadly anchored in the company among all employees.

A second hurdle in implementing Lean is applying its methods without fully understanding their details and interactions. As a result, you often stumble over small but important issues when implementing individual methods. Successful Lean implementation requires familiarity with the content of the methods.

If these two hurdles are not consciously addressed, Lean will not be successful. Therefore, the transformation to an excellent company is then called into question. This is unfortunate, as Lean production is a proven approach that can lead to operational excellence and drive a company's success.

Our exercise book aims to enhance how companies engage with Lean across all hierarchical levels, elevating application-oriented methodological expertise. We believe that this preparatory "doing" will promote the understanding and competence for Lean, making Lean transformation successful. We have developed 35 exercises focused on "doing Lean", while combining practical application with reflective learning. These exercises mostly relate to an application in the fictitious company LeanClean Inc. In this way, we create a concrete reference to "real world problems" and the understanding of the methods is greatly enhanced by its practical relevance.

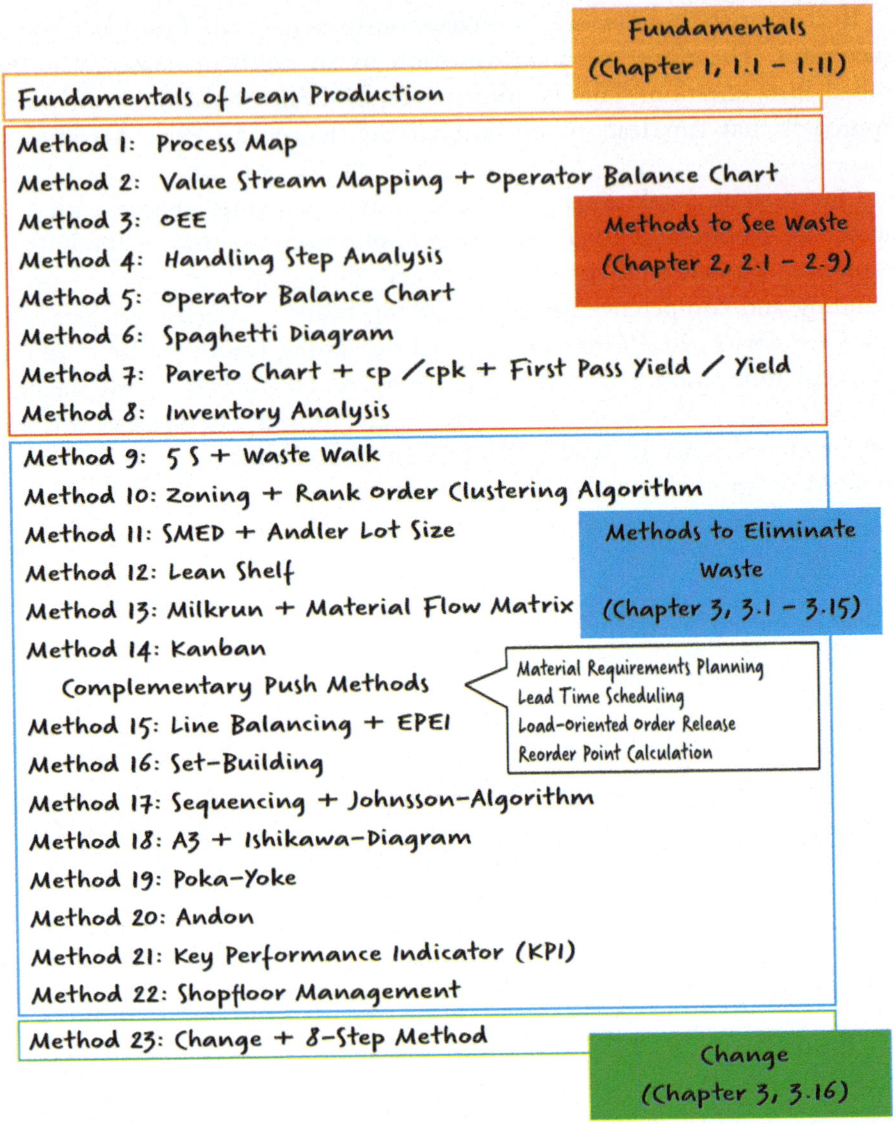


We begin our workbook in Chap. 1 with a brief summary of Lean fundamentals, focusing on the "seven types of waste" and the "nine principles of Lean". This is just a brief outline of the Lean theory. For further content and explanations of Lean in general, Lean principles, as well as detailed descriptions of the methods, please refer to the book *LEAN Production—Simple and Comprehensive*

In Chap. 2, you will find examples of exercises on methods to identify waste (Sects. 2.2–2.9). In Chap. 3, we focus on exercises to eliminate waste (sects. 3.1–3.15). In the final exercise section (Sect. 3.16), we provide examples to demonstrate how to implement Lean comprehensively. After all, how you lead, and manage change is crucial to the success of Lean.

We wish you enjoyable practice sessions, and we are confident that you will achieve expert status as a Lean professional.

# Contents

# About the Authors

**Roman Hänggi Professor of Production Management**

After studying engineering and earning a doctorate in economics, Roman began his career in the industry in the early 1990s by optimizing an optics manufacturing facility at Leica with Lean. His enthusiasm for Lean has accompanied him throughout his professional life. He implemented further successful Lean projects in production at Bosch, Hilti, and Arbonia and leveraged this production experience to optimize processes in development, service, and administration with Lean. His curiosity and desire to share his extensive industry knowledge led him to the Chair of Production Management at the University of Applied Sciences OST. There, he motivates students in Rapperswil and St. Gallen with lectures on Lean and digitalization in industry. Committed to practice application, Roman supports industrial companies in Switzerland and abroad in their journey to become Lean Champions and Industry 4.0 winners. He is also a lecturer in executive programs at the University of St. Gallen (HSG). In his free time, Roman enjoys skiing on the slopes in Appenzell or playing golf.

**Patrick Balve Professor of Logistics Planning and Organization**
After studying mechanical engineering with a focus on project management, Patrick worked in research and management consulting at the Fraunhofer Institute for Manufacturing Engineering and Automation (IPA). There, he supported numerous companies in improving their order fulfillment processes and organizational structures. As part of his doctoral thesis, Patrick researched the transfer of management cybernetics principles to improve the adaptability of manufacturing companies. During his subsequent eight years as a specialist and manager at a major German automotive manufacturer, Patrick deepened his expertise in Lean implementation with a focus on logistics and quality management. Since 2009, Patrick has been sharing his knowledge with students at Heilbronn University of Applied Sciences (HHN). Here, he inspires students of production engineering and mechanical engineering with Lean methods, principles of Lean logistics, and the application of agile and traditional project management. It is always important to him that the benefits of Lean are directly experienced through practical examples and exercises. To this end, he also manages the Learning Factory jumpING at Heilbronn University, a learning laboratory for the project-oriented and holistic experience of the product creation process.

**Lukas Budde Post-Doc for Production Management**
Lukas Budde studied mechanical engineering at the Technical University of Darmstadt and received his Ph.D. from the University of St. Gallen. Since 2015, he has been a postdoctoral fellow and deputy head of the Production Management Department at the Institute of Technology Management at the University of St. Gallen. He is also a member of the management team at the Complexity Management Academy in Aachen. His research focuses on smart manufacturing, operational excellence, and digital platform ecosystems in manufacturing companies. Lukas has supported companies in many digital transformation and production optimization projects. In addition, Lukas leads several international research projects in the field of production management. He is a lecturer for operations management and for customized programs at the Executive School of the University of St. Gallen. Since 2018, he has been the managing director of INOS, the largest regional innovation network for SMEs in Switzerland.

# 1

# Basics of Lean Production

## Contents



## 1.1 Why Do We Differentiate Between Waste?

### Everything Starts with an Understanding of Waste

Lean manufacturing is not a new concept. Originally developed in Japan some 60 years ago, the theory has evolved over time. Many companies have realized the potential of Lean management. However, experience has shown that the complexity of Lean is not in understanding the theory, but in practicing it. This process is often seen as difficult and unfamiliar because Lean requires a shift in mindset and a gradual, step-by-step implementation

© The Author(s), under exclusive license to Springer-Verlag GmbH, DE, part of Springer Nature 2025
R. Hänggi et al., *LEAN Production Training for Practice and Study—35 Exercises with Solutions*, https://doi.org/10.1007/978-3-662-70468-4_1

involving both employees and management. This is where we want to start and focus on action through practice.

Lean production is particularly important in todays's digital age. We want to digitize stable and waste-free processes, as this is the only way to ensure that significant investments in digital approaches are worthwhile. This is why we advocate for the sequence: "Lean first" and "digitalization second". While this approach isn't absolute, it emphasizes a mindset that begins with a thorough understanding of waste.

Since Lean is both the basis and the prerequisite for digitalization, its importance is significantly increasing, especially in today's world. Operational excellence can only be achieved by combining Lean and digital. Moreover, digitalization can elevate various Lean approaches to a new level. For example, it makes updating KPIs (key performance indicators) on a shopfloor board easier, and it can also simplify and enhance the transparency of information flow in Kanban processes.

We now understand that Lean is essential for achieving operational excellence in the digital age. But before diving into exercises, it's important to cover some Lean theory first—just enough to understand the basics of Lean manufacturing.

Any optimization using the Lean management approach starts with understanding of waste. The largest part of any process often consists of waste. It is not uncommon for a process to have 75% waste, leaving only 25% value added. In other words: "You're doing a lot of useless stuff in this process".

By "value-added", we mean anything the customer is willing to pay for. For example, they certainly don't pay to go to the warehouse four times to pick up handles for a production order for the Elephant vacuum cleaner. In Lean theory, waste is commonly categorized into different types, which we refer to as "waste types". This characterization helps to find, quantify, and ultimately eliminate waste in processes.

Understanding waste and its specific characteristics is the 1 × 1 of Lean management because everything is based on it. So, let closely examine these 7 types of waste. Once you understand them using some practice, you can spot them in everyday situations. We have also prepared a special exercise for you to try out at home later.

The Seven Types of Waste:
1. Overproduction
2. Stock
3. Transport
4. Motion
5. Waiting
6. Unnessary processes
7. Scrap and rework.

Now let's discuss waste using everyday examples.

## 1.2 Waste 1: Overproduction

### Too Much and Too Soon

You invited some friends over and planned to serve spaghetti. To ensure everyone had enough, you cooked a package. However, you ended up drinking more than you ate, and now there are twelve servings left in the pot. This overproduction is particularly unfortunate because it automatically leads to additional waste. Now, you need to put the leftover spaghetti into storage cans, store them in the refrigerator, and remember they are there. Over the next few days, you can use five servings for microwave lunches, but the rest will go bad and end up in the trash.

### It Is Still Waste

Whenever you produce something earlier or in larger quantities than needed, it results in overproduction, which is considered waste.

But is producing more always wasteful? For instance, it doesn't pay to pick three strawberries for every slice of bread, cut them up, and boil them with sugar for an hour. Wouldn't it be more efficient to make a whole jar of jam, even if it means overproduction? Certainly, it makes sense and is cheaper. However, you must be careful not to produce too many jars of jam ahead of time that end up rotting in the cellar. The key is to identify waste in a process first. You might not immediately think of a solution to streamline the process, but consistently identifying and acknowledging waste can encourage you to seek and discuss better methods.

## 1.3 Waste 2: Stock

### Stock Gives a Sense of Security

As a squirrel, you have the foresight to stock up on nuts to ensure you get your caloric intake, even if the "hazel bush" supplier fails to deliver during the the winter.

As this camp needs space, you must bury the nuts and remember where they are buried. When you retrieve the goods, you must first locate the nut. And when the food is finally found and dug up, you discover that the expensive food is now rancid or completely rotten.

While warehouses provide the comforting feeling of a secure supply, they also create many additional problems, such as inconsistent and uncertain delivery. The "hazel bush" simply needs to deliver more reliably—evolution still has some work to do!

In today's world of supply chain disruptions, a lack of inventory, or at least significantly reduced inventory, is often blamed for poor delivery performance. In fact, the opposite is true. If you are experiencing supply constraints, there are many reasons causing it. Knowing the real problems is important and critical to avoid surprise. The problem is not a lack of inventory, but unstable processes and unfavorable supplier partnerships. Lean is not the cause of the supply problem, but rather the failure to apply Lean comprehensively!

### The Money Is Not Only Tied Up in the Goods!

When screws, motors, or food are stored, money is tied up in goods that could be put to better use than lying around. But the money tied up is usually not even the most expensive part of the inventory.

You should not underestimate the effort required to build and operate the necessary warehouse: you need shelves, pallets, and containers. Forklifts are required to move items around, and computer programs manage inventory and storage locations. There is also work involved in storing and retrieving the material at the right time, which requires hiring and training people. We count the material for the inventory. And since the entire operation cannot occur outdoors, a building is needed—one that must be heated in the winter and cooled in the summer. All these factors make storage much more expensive than the inventory value would suggest.

Repeatedly, we see attempts to automate warehousing to manage inventory and reduce costs. Fully automated warehouses are the ideal solution. However, the reality is that automated waste is still waste. It is also often forgotten that even automated storage and retrieval is time consuming and should not be underestimated.

## 1.4 Waste 3: Transport

### Parts Tourism

Transportation always occurs because process steps are physically separated—and transportation is wasteful. Consider the production of a T-shirt is a perfect example. Cotton is grown and harvested in the U.S., then shipped to China for several weeks. In China, the thread is spun, transported by truck, and then transported again for several days. The fabric is woven and shipped again. Finally, the T-shirt is sewn, packaged, and shipped halfway around the world to Germany, where it awaits buyers in the store. As with the other types of wastes, there is always a reason for the waste.

**Whether a Few Meters or Halfway Around the World, Both Are Transport**

In T-shirt production, it may be economically advantageous to produce in low-wage countries. However, the transportation involved in this process is wasteful.

Even short-distance transports require organization and coordination. These add complexity to the process: What is to be transported? From where and to where? Who carries out the transportation? By what means? The answers are containers, vehicles, drivers, accompanying documents, computer, and tracking systems. You need cranes, forklift trucks or personnel to load and unload the goods. In summary, transportation incurs significant costs, consumes large amounts of energy, and often does not enhance the quality of the goods for the customer. This clearly exemplifies waste.

## 1.5 Waste 4: Motion

### Even Short Distances Lead to Espresso

All paths and movements required to add value are considered wasteful, because even necessary paths are waste. For example, the distances your feet or hands must travel to reach a cup of coffee are waste.

It is also important to consider how often you need to walk that distance a day. Frequently used utensils and ingredients, such as coffee cups, naturally weigh more on the usage scale compared to less frequently needed items like water or refill bags of coffee beans.

### Many Few Also Make a Lot

After all, as the saying goes, even small things make a big difference and many small steps make a marathon. It's two meters from the cupboard to the coffee machine, meaning four meters there and back. Over the course of a day, with about 300 cups of coffee brewed, that adds up to 1200 m. In a month, our barista covers almost a marathon distance. Over a year, that's more than 400 km!

So, which is more time-consuming: moving the cabinet one meter to the right or moving from Frankfurt to Hamburg?

## 1.6 Waste 5: Waiting

### If It Takes a Little Longer…

Nobody likes waiting. After all, who would deliberately choose the longest line at the supermarket checkout? From Lean perspective, waiting is seen as a form of waste and undesirable. For example, our barista has to wait 25 s for the coffee to flow into the cup, and the coffee didn't improve with time.

While 25 s may seem like a short interval, it adds up to several hours over the course of making 300 espressos a day. And this waiting wasn't even relaxing.

## 1.7 Waste 6: Unnecessary Processes

### Newsletter and Lost Glasses

Everyday, life is full of unnecessary processes. Think of finding your glasses or keys, opening and closing cupboard doors in the kitchen, fiddling with stickers on an apple, deleting the thousands of newsletters in your email account, or—always fun—picking pins out of a new shirt (only to forget one, ouch!).

In most cases, we have become so accustomed to these unnecessary processes that we do not even think about the effort and time wasted. Even when we do, it's easy to make excuses: "That's just the way it is! There's nothing you can do about it! I don't have time to change anything right now". However, it's worth the effort to change the process in the long run. So, take a moment and unsubscribe from those unnecessary advertising newsletters!

## 1.8 Waste 7: Scrap and Rework

### It Does Not Always Succeed the First Time

Things don't always go perfectly the first time. The phone has rung, the oven has been forgotten, and in the end, it has mutated into a smoker grill. Under certain circumstances, the marble cake can still be reworked and saved with an inch-thick chocolate icing. However, if the phone call took too long and there is more charcoal than cake in the pan, the work will end up in the trash—and money, time, and sweat will be lost along with the cake.

Fortunately, you will notice a charred cake. If the sugar is missing, you will only notice it when you eat it—right in front of your whole family! How embarrassing! The key question is, how to ensure the "waste" doesn't happen next time!

## 1.9 Seeing and Understanding Waste

**Yes, but…**

Yes, but storage is necessary! Or perhaps you're thinking, "it is much more efficient to produce in large quantities". We understand that this strict categorization of waste might feel challenging, but we encourage you to simply name the waste in your process. Even if you can't immediately think of a way to reduce inventory, cut waiting times, or shorten transportation routes, the first thing to do now is to make the waste transparent.

By daring to name the waste and question existing processes, you will begin to see them in a different perspective. Hidden problems will then become obvious.

**Sometimes You Have to Look Closely…**

Not everything that looks like waste is truly waste. Take wine, for example: when it is stored in a cellar and gains value as it ages, the storage is not considered wasteful inventory; it's a value-adding process. Some jobs, such as lifeguard, are all about "waiting". Lifeguards may sit motionless while they watch the water, peeking for drowning people. If the product is called "safe bathing", this activity is important because it ensures safety, but from Lean perspective, it is still wasteful. Even a seemingly necessary process, such as monitoring a critical system in production, is wasteful. Inspections, although arguably necessary for quality or safety reasons, are always wasteful.

## 1.10 Understanding of Lean Principles Is the Foundation

To combat waste, we use several Lean principles for optimization. These principles aren't implemented directly but rather provide a way of thinking about how to make processes lean and waste-free. These principles are the foundation of all Lean methods. Let's examine the 9 principles of Lean manufacturing.

**Principle 1: Pull Principle**

The pull principle encourages you to think from the customer's perspective: you produce only what the customer wants. Most importantly, you don't stock products based speculative planning. While we acknowledge that this isn't always feasible, adhering to the principle of "thinking strictly from the customer's point of view" helps focus on customer demand rather than batch production according to a plan. The opposite, and unfortunately more common, principle is the push principle, which involves forecasting demand and producing goods based on a production plan for stock.

**Principle 2: Flow Principle**

The flow principle is closely related to the pull principle. It dictates that the products or parts in production must flow, ensuring that no goods stagnate between processes by organizing the sequence of your production steps. This is often referred to as just-in-time (JIT) production, especially when the flow principle aligns with current customer demand. To implement the flow principle effectively, you must observe the setup times, as these directly determine batch sizes.

## Principle 3: Takt Principle

If several processes are to be conducted in succession, the takt principle leads to a flow between the work processes. Only when all processes follow a uniform rhythm, waste, such as product backlogs between processes, can be eliminated. The takt principle aligns all work processes with the customer's demand, generating a flow between the work processes. The flow principle and takt principle form a logical unit. Takt brings flow and flow simplifies the takt. This close relationship often leads to confusion between the two in practice.

## Principle 4: 0-Defect Principle

To eliminate the waste associated with "scrap and rework", you must break the chain of defects. This implies that defects are neither accepted, created, nor passed on. While the zero-defect principle sounds logical and simple, implementing it seriously requires radical thinking. It's easier to order a few extra parts—just in case. And who has the time to analyze and eliminate the source of the mistake in the midst of daily operations? However, the challenge lies in rigorously identifying and eliminating the root causes of defects.

**Principle 5: Seperating of Waste and Value Creation**

Every second counts in a complicated operation. For example, if a surgeon needs a scalpel (customer benefit = value creation), fortunately, he should not have to search for one in storage. Assistants are there to provide immediate support, eliminating unnecessary movements (paths = waste). Fortunately, the processes of value creation and waste are separated. By the way, this also applies to production—production workers should not waste time searching for parts in the warehouse, as this would disrupt the production process—logistics should handle that.

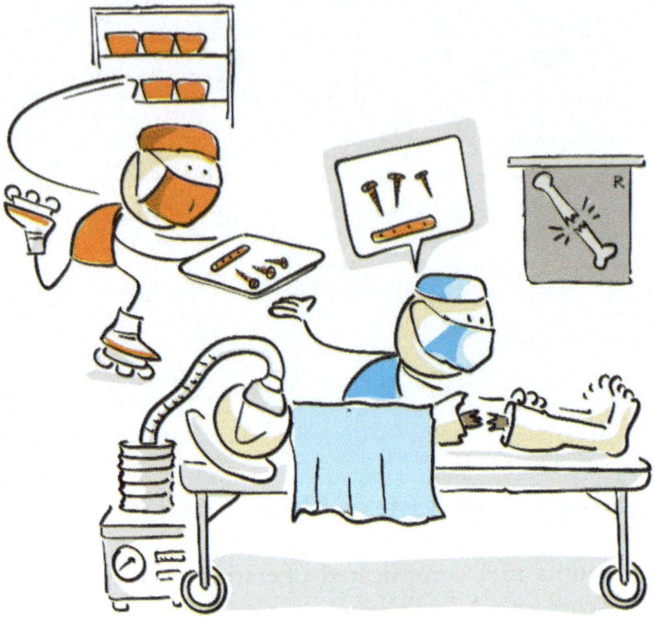

## Principle 6: FIFO Principle

Think about the first-in-first-out (FIFO) principle with yoghurt in your refrigerator at home. Do you always place the newly bought yoghurt behind the older ones? This practice ensures that you consume the "older" yogurts first and the "newer" ones later. The FIFO principle must be applied to all storage processes to prevent waste. This also applies to circulating stock and buffers in production.

### Principle 7: Minimum Distances

Unfortunately, people often take the shortest route, which is often not the case in the factory and in many other areas. As a result, in our experience, walking is a significant time-waster in many processes.

When designing processes, it is crucial to consciously and "in principle" consider how to minimize distances, not just for walking but also for reaching distances and every other movement. For example, placing a towel next to the bathtub not only minimizes distances but also reduces the amount of cleaning in the bathroom (an unnecessary process), which reduces waste, twice.

### Principle 8: Value Stream Orientation

Value stream mapping involves examining the entire process chain, not just from one isolated process, but from the customer order to the product delivery, which constitutes the value stream.

Consider the process of waste disposal: for example, in the individual process, while disposing of all waste unsorted is easier, separating the garbage would be time-consuming. The problem is that separating the waste into two further process steps is disproportionately more costly. Therefore, it is important to consider the overall effect whenever a change is made.

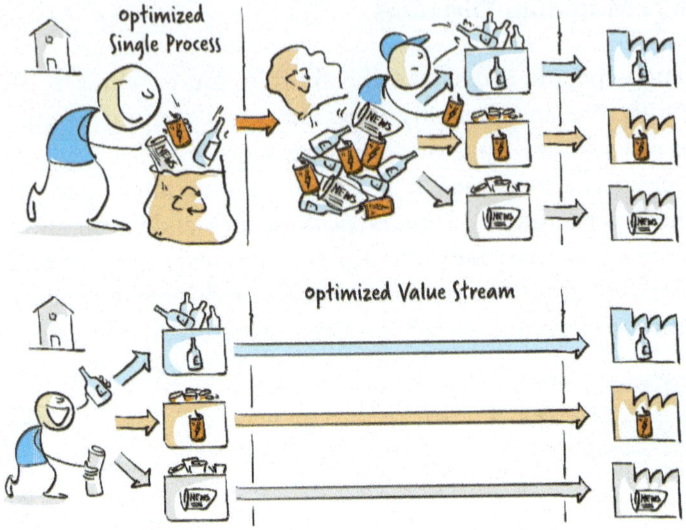


## Principle 9: Standardization

We know the problem from everyday life. We search for our glasses because we left them somewhere or forgot them, and eventually find them on our forehead. Or we store five different cell phone chargers in a drawer because the plug has been changed for each new device. Such situations illustrate the importance of standardization to eliminate waste. Making standardization one of the 9 Principles underlines its importance and significance in implementing Lean management effectively.

Incorporating standards into processes helps prevent mistakes and make work easier. These standards should be reinforced through training and their effectiveness verified through audits.

## 1.11 Now It Is Your Turn

We are now familiar with the 7 types of waste and various Lean principles. To put this knowledge into practice, we will start by identifying waste and then, apply the Lean principles to reduce it using various methods presented in Chaps. 2 and 3. However, you can start practicing now. Waste can be found everywhere—at home, in shopping mall, on your way to work, or at the gym. Try to develop "eyes for waste" by consciously looking for it. Then, think about how you can eliminate or reduce it.

But if you've gotten used to the waste because it's always been there, you may not see it anymore. It has somehow become the norm. There's a simple trick to seeing waste again: Take a different route to your workplace, consciously change your routine or workflow. These changes will help you see the waste again because the process is new or the perspective on it is different.

Give it a try. This training is an excellent foundation for other Lean exercises.

# 2

# Exercise on Methods for Seeing Waste

## Contents



© The Author(s), under exclusive license to Springer-Verlag GmbH, DE, part of Springer Nature 2025



R. Hänggi et al., *LEAN Production Training for Practice and Study—35 Exercises with Solutions*, https://doi.org/10.1007/978-3-662-70468-4_2

## 2.1 Welcome to LeanClean Inc.

For most of the exercises, we use cases from LeanClean Inc., a vacuum cleaner manufacturer. LeanClean Inc. is a company known for its great products, but its somewhat dusty production and assembly has grown organically over the past 30 years. Over time, the company has pragmatically improved individual production processes and optimized workplaces here and there, always showing great commitment, but with little system and strategy. After a generational change, a fresh breeze is blowing through the boardroom at LeanClean Inc. The management now wants to take the "Lean" in the name a little more seriously.

LeanClean Inc. manufactures vacuum cleaners in three product lines, each designed to meet specific customer needs. The best-selling LeanClean product is the entry-level Snake, a solid standard vacuum cleaner that is easy to use and, above all, very attractively priced. It has no prestige, but as a cash cow, it fills the coffers of LeanClean Inc. In the last few years, production has steadily increased to 150,000 units/year. However, it is expected to at least double in the next 3 years, as a new generation has been introduced. This new generation is even more cost effective, and the dust-removal performance has been improved, making the Snake suitable for every household. This aligns with the current consumer preference for products that are "good enough at a fair price". Another interesting feature of this model is the recycling percentage of the plastic used. Almost 80% of the components are made from recycled material that focus on the circular economy business model, giving vacuum cleaner an extra "boost".

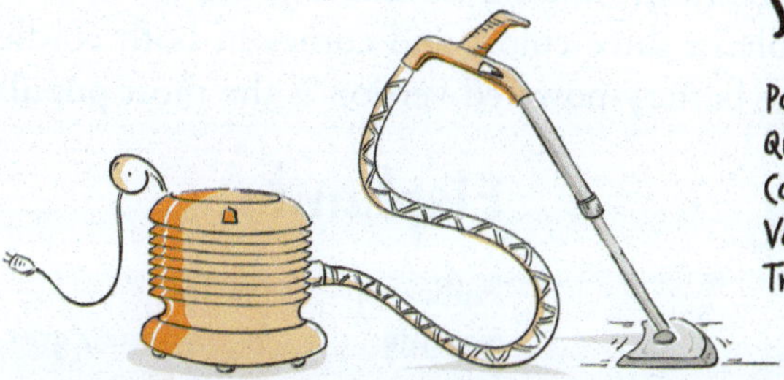

## Snake

Positioning: Cheap line
Quantity: 150,000 pieces/year
Colors:
Variants: None, only one model
Trends: High cost pressure

The Octopus is designed to be an all-rounder. In addition to vacuuming, it can iron shirts, wipe windows, and tidy up. In fact, you can do almost anything with this vacuum cleaner. With its five cleaning programs, it solves any dirt problem. The Octopus is currently only available in orange, but customers will soon enjoy it in many other colors. While this comes at a high price, it hasn't deterred its large fan base from buying it. To keep these tech-savvy early adopters satisfied, LeanClean Inc. plans to introduce even more groundbreaking features next year, though these developments are still being kept secret. The Octopus has so many features because there is a constant demand for "specials" in small batch sizes. This demand has implications for LeanClean Inc., as discussed in Sect. 3.7, where we explore how a new organizational approach requires an adapted production system.

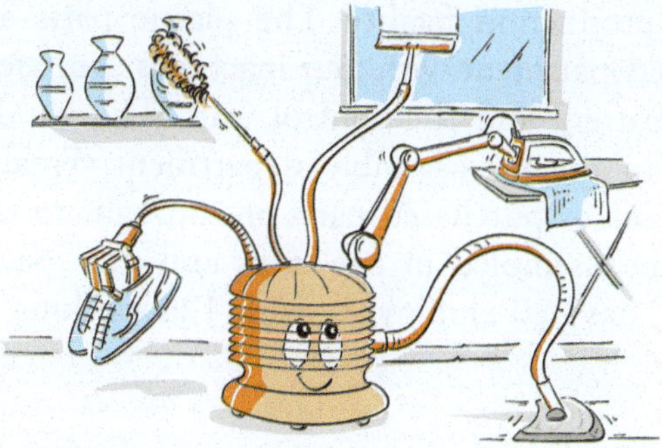

## Octopus

Positioning: All-in-one solution
Quantity: 20,000 pieces/year
Colors: (More colors soon)
Variants: Additional features in planning
Trends: New models and options every six months

Finally, we come to the Elephant, the latest innovation in LeanClean Inc.'s portfolio: this model is available in three colors and comes in both cordless and corded version. The orange battery-powered version is the most popular.

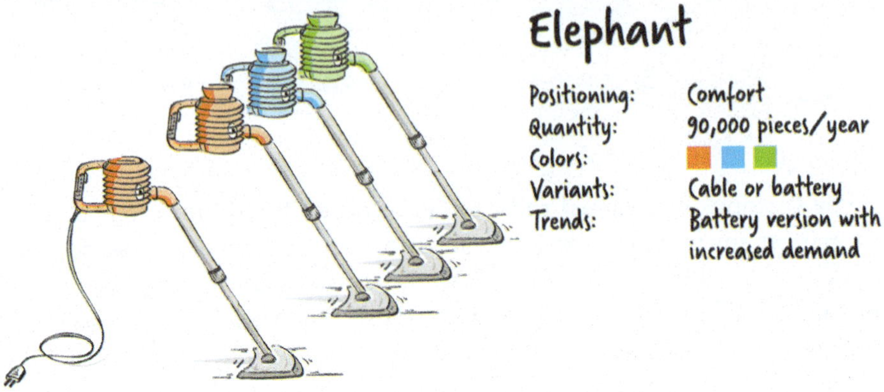


The Elephant's success has exceeded the boldest forecasts: sales have increased from 50,000 in the first year to 90,000 in the second year, since its launch. The orange Elephant is the best seller, with nearly 200 units sold per day on average, followed by the blue and green models, which each sell around 100 units daily.

A factory tour offers an initial insight into the production processes at LeanClean Inc. All three vaccum cleaner models and their components are manufactured in the LeanClean production facility. The plastic parts are produced in the injection molding department with four machines. The electronics department supplies the battery and the control unit. The handle shells and switch are assembled in the pre-assembly department. Finally, all in-house components and dozens of purchased parts are brought to the final assembly area, where they are assembled in five steps and then packaged for shipping. LeanClean Inc. has 220 employees, with 129 working in production.

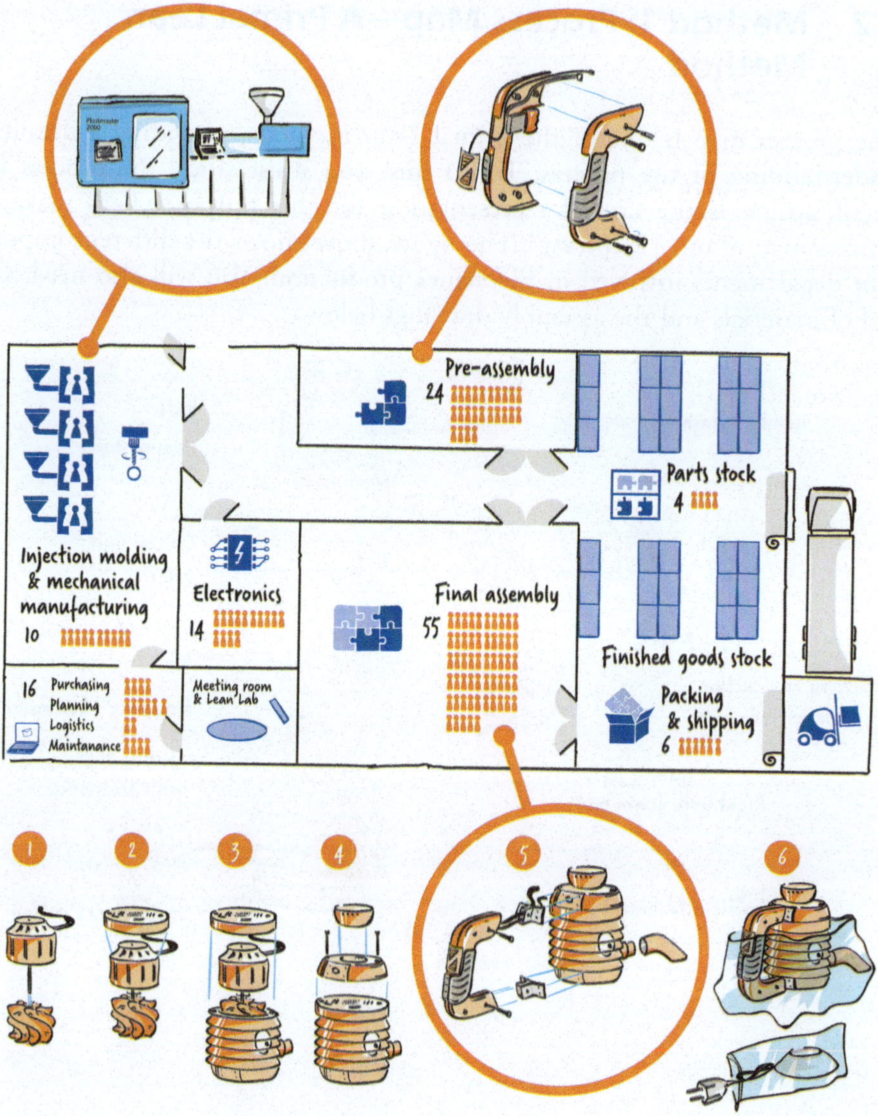


LeanClean Inc. is our factory for applying these methods, which will help the company become more productive.

## 2.2 Method 1: Process Map—A Primal Lean Method

The process map is one of the primal Lean methods, providing a detailed understanding of the process. To do this, you must study the process in detail, which is the crucial representation for this purpose. Let's make a process map of our "Elephant". The figure above shows the different important departments involved in Elephant's production. You will also need the bill of materials and the assembly drawings below.

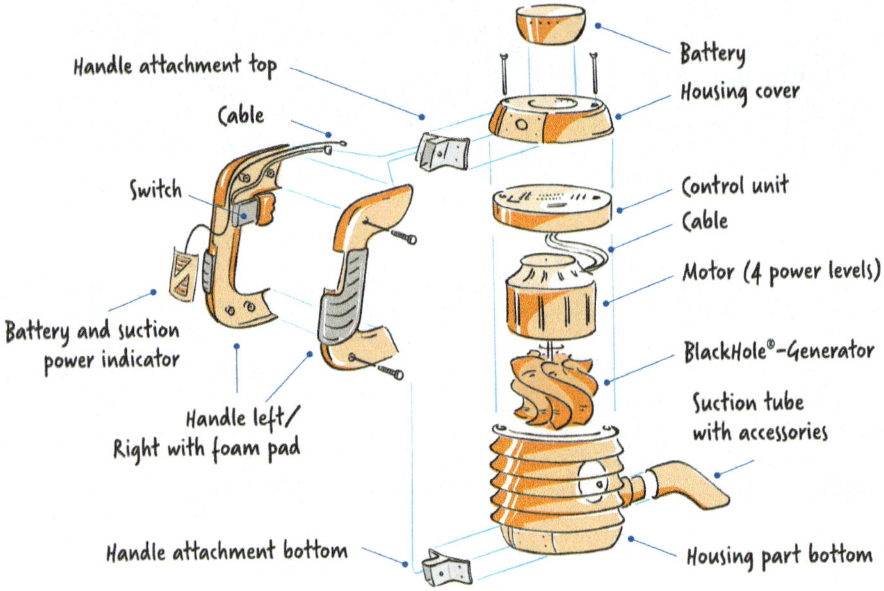

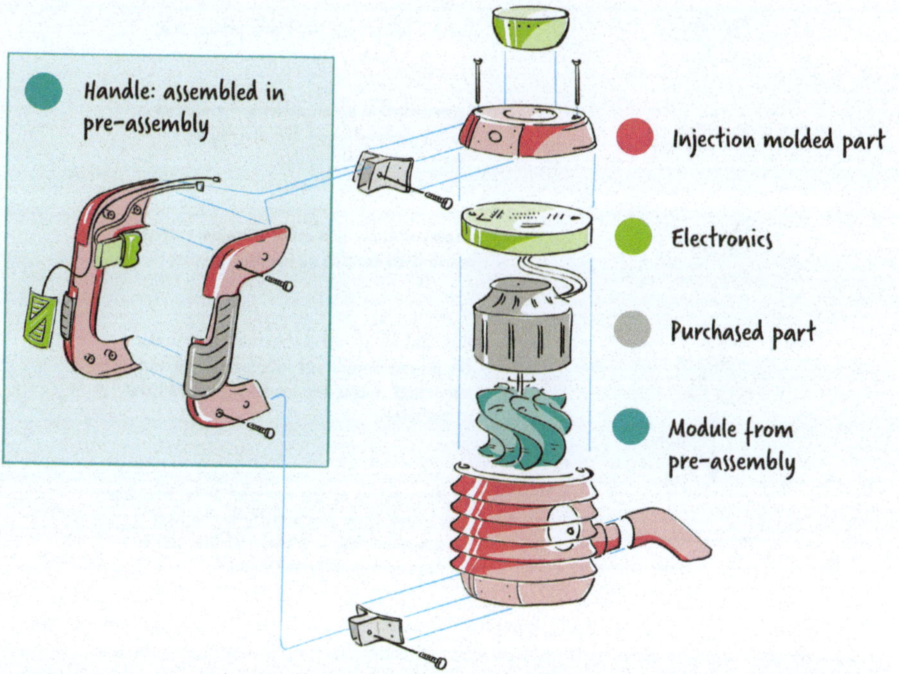


All right! Now, you are ready to create a process map for the Elephant's production using the swim lane representation. As you do so, keep the following hints in mind:

**Tip** Symbols: Use few symbols. Again, simplicity is important for understanding. The following symbols have worked well in practice:

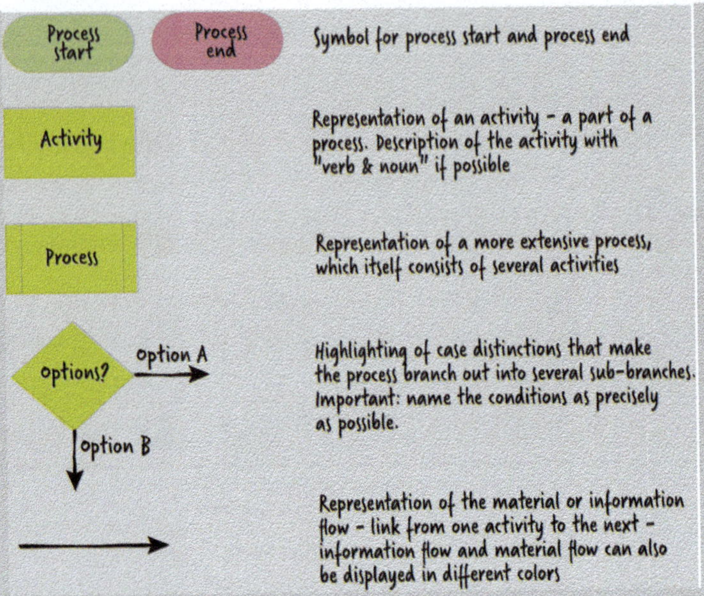


In practice, no distinction is often made between activity and process. Strictly speaking, an activity is a single action within a process, whereas the process comprises the sum of these activities. If you prefer not to differentiate, it is best to use the "simple rectangle" of activity for process and activity alike. This is the easiest because it is immediately understood.

A swimlane diagram is a structured form of flowchart. Typically, the process is often plotted along the horizontal (x-axis) and the vertical (y-axis) is organized by department into swimlanes. The process flows from left to right within these swimlanes.

The best way to begin a current state map (and this exercise) is to list all the departments involved or visually highlight them in the text. Use this information to create a first draft of the swimlanes. Now you are ready to enter the first process symbols. If at any point the diagram becomes too confusing, arrange the horizontal department swimlanes to achieve a clearer layout.

Understanding any process is greatly enhanced by its graphical representation. It is also best to establish this process in a team. This way, the process is widely shared and discussed, which not only strengthens the common understanding, but also improves the quality by incorporating different perspectives.

A flowchart is never perfect on the first time. The secret lies in getting started! Start at a point that seems plausible to you and adjust from there until you have a coherent and logical sequence of activities across multiple departments. To facilitate changes, we recommend working on a pinboard or a whiteboard. If the team opts for direct digital documentation, ensure that only a suitable person is responsible for managing the software.

Keep the description of activities in the process map at a consistent level of detail. If you need to go into more detail, use the "process symbol" to indicate this.

After assessing the current situation, focus on identifying waste. Use a waste flash to highlight possible wastes. It is also helpful to add the type of waste to the waste flash—this will make your statement more specific.

**Tip** The future-state processes are the result of a comprehensive analysis of the current situation and an examination of the waste. Now that the team has a comprehensive and shared understanding of the current process and waste, they can work together to define a well-founded target process.

The handle consists of two handle shells that are made in the injection molding department. The granules required for this are supplied from the warehouse. Other handle components, such as the cable, switch, and display are purchased parts also stored in the warehouse. The cable is soldered to the switch and display in the electronics department and then returned to the warehouse as a "switch/display" sub-assembly. At a later stage, this sub-assembly is removed from inventory and combined with the previously stored handle shells in the pre-assembly department to form the handle. The finished handle is then returned to storage. The process is finalized during the Elephant's assembly (assembly of cover, control unit, handle, etc.), where the handle is then incorporated into the final product.

In its current state, LeanClean Inc. generally produces everything to stock ("make to stock"), i.e., every completed production order for individual parts, assemblies, and finished products (based on the forecast of products to be sold) always goes directly to the warehouse. However, there are instances where a customer order directly triggers a complete production order ("make to order"). In these cases, the fully assembled handles are delivered directly to final assembly, and the finished Elephant is dispatched straight to the shipping department, bypassing the warehouse.

***Task***

1. *Depict the detailed process described above "Injection molding and handle assembly" (components outlined in turquoise in the assembly drawing) in a horizontal swimlane diagram. Show both, the "Make to stock" process and the "Make to order" process, in a joint process map.*
2. *Now we want to focus on the logistics and administrative processes. Can you identify potential improvements in logistics and the warehouse processes? It might be useful to look at the exercises on Kanban in Sect. 3.6 first (just as broad hint).*

***Solution***

1. *The following possible process map details the assembly procedure for the handle shells at LeanClean. The handle shells are first manufactured by the*

*injection molding department and then delivered to the warehouse ("make-to-stock"). The purchased parts—such as cable, switch, and display—are taken from stock and delivered to the electronics department where they are soldered together to form a "switch/display" sub-assembly. This sub-assembly is then returned to storage. Next, the handle shells and the switch/display subassembly are then removed from storage and sent to the pre-assembly department. Here, all the parts are assembled into the finished handle, which is subsequently stored as the handle assembly. However, there is also a case where the handle shells and this handle can be delivered directly to assembly—whenever a a customer order ("make-to-order") from an Elephant directly triggers a production order. In this case, the handles are not stored, but delivered directly to the assembly line.*

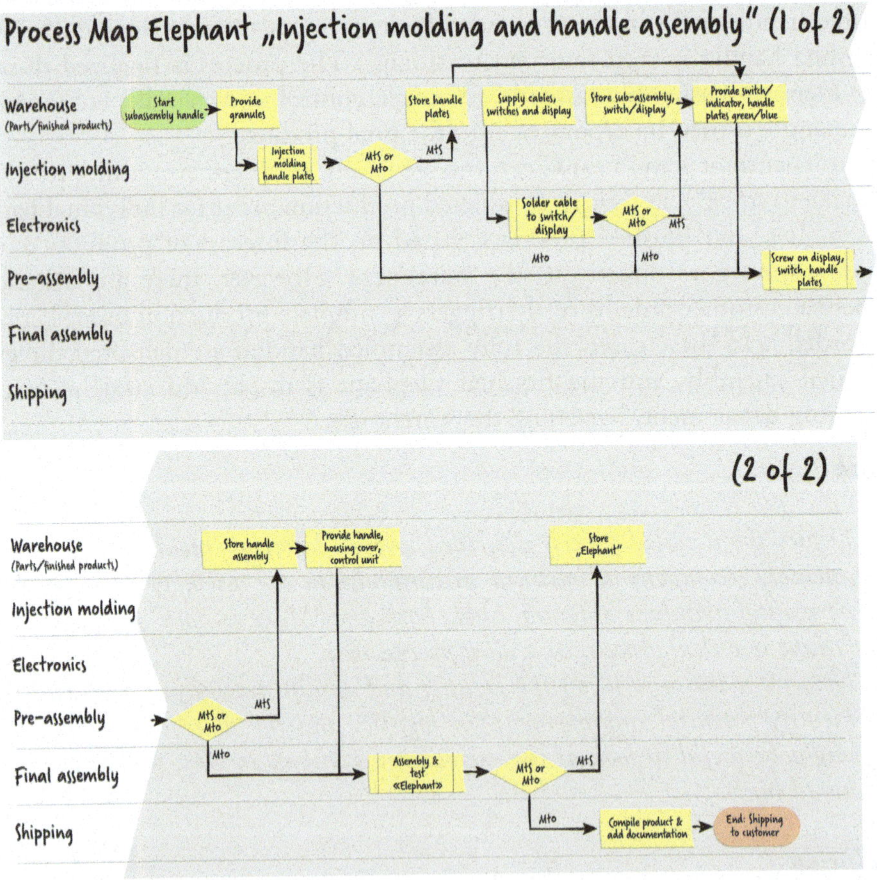


2. *The introduction of Kanban control loops significantly reduces the complexity of the process. You will eliminate many processes by reducing inventory; you*

*will also massively reduce throughput time, reduce complexity, increase transparency, reduce inventory costs, and also produce only what you actually need. If there are no new orders for the Elephants, you won't produce to stock. This approach represents a substantial improvement for LeanClean Inc., aligning production with demand and optimizing resources.*

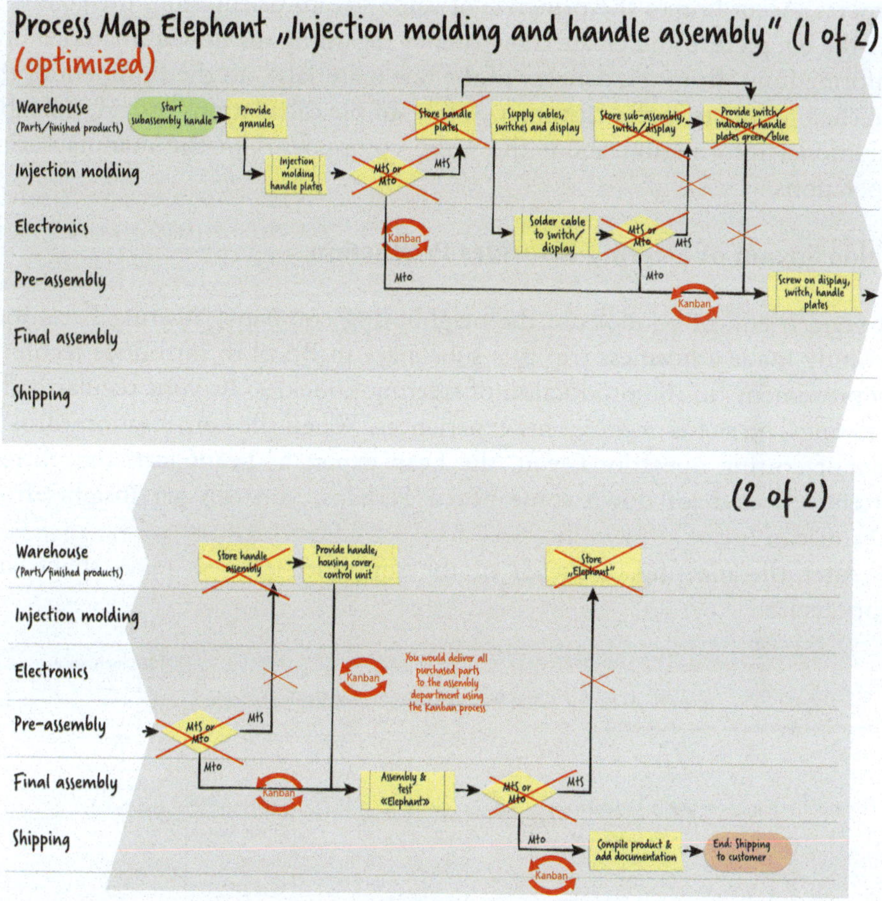


*Only by visualizing the entire process in a process map, can you identify these areas for improvement. This approach enables not just you, but your entire team, to see the process clearly. As a result, the decision to make a change is based on facts and supported by everyone. Another way to enhance transparency of the current situation is to use value stream mapping (see the next exercise in Sect. 2.3).*

## 2.3 Method 2: Value Stream Mapping for Comprehensively Analyzing the Material Flow

Value stream mapping provides a bird's-eye view of the manufacturing process. According to the principle of value stream orientation, the goal is to see the big picture and create transparency regarding the flow of materials and information—from the delivery of the raw materials—to the shipment of the finished products to the customer. A few simple calculations provide valuable additional information about the waste components in the manufacturing operations.

### Value Stream of Steering Knuckles Production

A good friend of yours from the neighboring company, Steering Parts Inc., recently made a business trip to a subsidiary in Brazil to introduce technical improvements in the production of steering knuckles. At your regular coffee meetings, he shares some of his experiences. When the conversation turns to manufacturing operations, you, the Lean expert at LeanClean Inc., listens attentively and jot down some notes. Perhaps, you can get insights from LeanClean Inc.

After the meeting, a sketch of the simplified process takes shape and appears like:

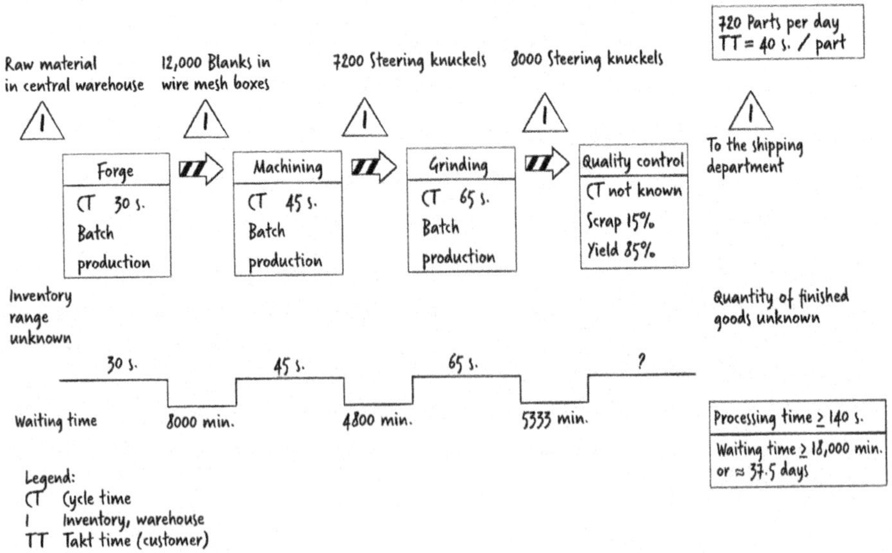

Although your friend has developed effective solutions for the technical challenges in the manufacturing process—the main reason for his trip—issues remain with delivery capacity and frequent overtime in many areas. The demand for parts is high, with 14,400 units required per month for three different customer plants in South America and Mexico. To better understand these challenges, you do a few simple calculations:

***Task***

1. *First, determine the average number of steering knuckles that need to leave the factory each day to meet the customers' requirements. Then, calculate the customer takt time (TT) using the usual formula:*

$$Customer\ Takt\ Time(TT) = \frac{working\ time}{customer\ demand}$$

> **Tip** Certainly, you must ensure that you constantly use the same reference value. If you take the customer demand per day, then the working time must also relate to a day. Alternatively, you can base all figures on a full month.
>
> Regarding the unit of time itself, you have a choice: You either calculate in seconds or in minutes. In most cases, you will quickly see which makes more sense.

2. *Next, add the timeline to the value stream map. Since little is known about the actual processing times in the areas, start with the cycle times (CT). Determine the individual ranges of the material buffers between the work areas as follows:*

$$Waiting\ Time = number\ of\ workpieces \times customer\ takt\ time$$

> **Tip** If you are in a hurry to calculate the total waiting time, you can simply add up the entire inventory along the value stream (WIP) and divide it directly by the daily delivery quantity. This will immediately give you the idle or waiting time during which the parts are "in production" but are actually just sitting around:
>
> $$Waiting\ Time = \frac{\sum inventory\ counts}{daily\ deliveries}$$

*You now have a good overview of the process and its wasteful components. However, demonstrating the causes of the unsatisfactory delivery performance to your friend, requires another calculation:*

3. *Based on the bottleneck process, i.e. the area with the longest cycle time, the first step is to determine which output per month is realistic—so far the company sticks to its regular single-shift operation. In a second step, you also deduct the scrap from the output of the bottleneck process. What actual delivery quantity do you get?*
4. *What advice would you give your colleague now?*

## *Solution*

1. *Customer takt time*

$$TT = \frac{\left(20\frac{days}{month} \times 480\frac{minutes}{day}\right)}{14,000\frac{steering\,knuckles}{month}} = 0.67\frac{minutes}{steering\,knuckle}\,or\,40\frac{seconds}{part}$$

2. *Value stream map with timeline*

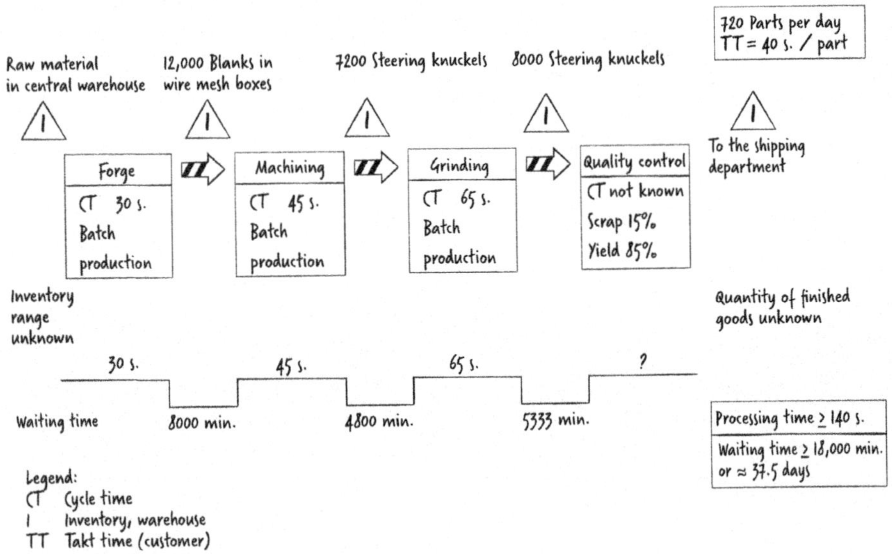

3. *The delivery quantity (Qty) is based on the bottleneck process "grinding", then reduced by the scrap rate in "quality control":*

$$Qty_{grinding} = \frac{monthly\,working\,time}{CT_{grinding}} = \frac{20\frac{days}{month} \times 28{,}000\frac{s}{day}}{65\frac{s}{part}} = 8862\frac{parts}{month}$$

$$Qty_{control} = Qty_{grinding} \times (1 - scrap) = 8862\frac{parts}{month} \times 0.85 = 7533\frac{parts}{month}$$

4. *Recommendations: Root cause analysis of the quality problems (e.g., with the A3 method, see Sect. 3.11), reduction of the cycle times in "grinding" to TT (– 10%), then also in "machining" (same issue CT > TT).*

## Value Stream of Riding Lawn Mower Production

Due to the recent bankruptcy of a competitor, Lawn Care LLC's lawn mower sales have doubled over the past year. No one anticipated this "happy event", and the sales department complains of persistent production backlogs. The order books are full for months ahead, with an average of 120 units per week of the popular LM180 model needing to be delivered. However, this target is far from being achieved.

Unfortunately, LeanClean Inc. is also among the waiting customers. The riding lawn mower is urgently needed for the orchard recently acquired as part of the "Healthier Eating" initiative. As Lean expert of LeanClean, you offer your assistance to Lawn Care LLC. Beyond achieving shorter delivery time with your assistance, you hope to gain insights with each new application of the value stream mapping method.

Fortunately, the main production facility is nearby, and after a quick phone call, an appointment with the plant manager was scheduled on short notice. Upon arrival, you learn that the LM180 mower causes the biggest challenges during the manufacturing process, with all other mower variants being of secondary importance.

Armed with a clipboard and under the expert guidance of the plant manager and a workforce representative, you begin exploring the LM180 production process. You start at the finished goods warehouse and move upstream along the production line to the receiving dock. You make the following notes during the interviews:

- The finished goods warehouse has 60 LM180 units, ready for pickup by the logistics provider.
- At workstation 4 "Test run": 2 employees, who are self-organized and always managing 2 riding mowers simultaneously. On average, one LM180 completes the test procedure every 20 minutes. This excludes the approximately 5% of all mowers that are sent to a separate workstation for more extensive rework. There are 12 LM180 units located in front of workstation 4 for the upcoming test run.
- Workstation 3, the "Final assembly" station, is staffed by 1 employee and has a cycle time of 40 minutes. In front of station 3, there are purchased parts and pre-assembled modules for 60 lawn mowers.
- Workstation 2, the "Pre-assembly" station, is staffed by 1 employee who performs the pre-assembly activities for one LM180 every 35 minutes. Setup times are negligible. There are purchased and in-house manufactured parts for approximately 48 lawn mowers, which are scattered throughout the pre-assembly area by logistics employees.
- Workstation 1, the "Cutting and welding" station, is operated by 1 worker who produces the frame parts, with a cycle time of 21 minutes ("when everything works", as they tell you). However, a maintenance worker reveals through phone call that due to the machines age, their average availability is only 80% and the management has already been warned severally about the need for a replacement investment. There are also quality problems with the spot-welding machine, leading to a reject rate of up to 25% on certain/no less critical parts.
- At first glance, the raw material for cutting the parts appears to be sufficient for about 10 working days (2 weeks).

### *Task*

*You thank them for the informative tour, and then you retire to the lounge for 15 minutes to evaluate the information you received in a few steps. Then, you answerr the following questions:*

1. *Determine the customer takt time (TT) based on a weekly 40 hours working schedule in a single-shift operation. In doing so, you have implicitly considered the occasionally reported overtime.*
2. *Create a value stream, focusing on the material flow, and include all relevant production information. Based on insights from the factory tour, all production areas receive weekly work schedules from the ERP system, which means that the material flow of the main components is based on the push principle. Finally,*

*you add a timeline, calculate the total processing time and the sum of all idle times (waiting times) per LM180 in days, and enter the waste flashes.*

3. *Using the information from the timeline, you can now determine the relationship between value-added time and waiting time in the production process. This metric is called "Degree of Flow" (DF):*

$$DF = \frac{\sum processing\ times}{\sum waiting\ times \times daily\ working\ time} \times 100\%$$

*Reflecting on the value stream analysis of the Elephant handle in your company, you recall that the processing time from the granules to the finished vacuum cleaner was 14 minutes—with a waiting or idle time of a whopping 53 days. To calculate the DF for the Elephant handle, consider the following:*

$$Degree\ of\ Flow(DF) = \frac{14\ minutes}{53\ days \times 900\frac{minutes}{day}} \times 100\% = 0.03\%$$

**Tip** It is important to remember that the processing time in the DF calculation includes both value-adding work steps and waste. For example, a closer examination at the Elephant final assembly revealed that the value-added share was less than 50%. In other words, the calculated degree of flow is still a sugarcoated value!

*Now, it's time to determine and discuss the degree of flow for the LM180.*

4. *Create the Operator Balance Chart (OBC, see also Sect. 2.6). Considering the required takt time, identify which stations is the bottleneck that slows down the entire work system. What short-term measures will you propose to the plant manager to ensure that the LM180 can be delivered to your company on time?*
5. *Two weeks later, while the situation has improved significantly, the sales department is unsatisfied with the delivery quantities and the plant manager is complaining about new overtime in the "Cutting and welding" department. As a result, you receive a phone call requesting another visit to Lawn Care LLC. As you prepare for the second visit, you realize that after eliminating the first bottleneck situation, the system availability and scrap rate in workstation 1 have become more significant and should have been considered by you right from onset. In reality, "Cutting and welding" behaves in the long run as if the*

*station had a much longer cycle time. This can easily be calculated as follows:*

$$CT_{effectiv} = \frac{CT}{availability \times yield}$$

*You now perform the same calculation for workstation 4 and enter both results in the OBC. What other improvements would you suggest?*

6. *To be better prepared for operational problems that inevitably arise, the work system should be designed for an operational takt time that is 10% below the customer takt. How can this goal be achieved in this situation?*

### *Solution*

1. *Customer takt time:*

$$TT = \frac{working\ time}{customer\ demand} = \frac{40\frac{hours}{week} \times 60\frac{minutes}{hour}}{120\frac{pieces}{week}} = 20\frac{minutes}{piece}$$

2. *Completed value stream map:*

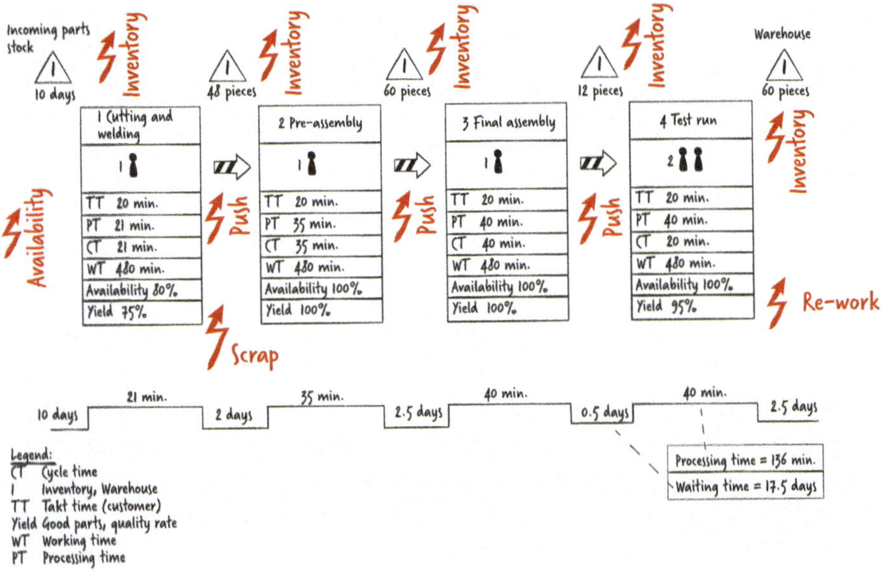


3. *Degree of Flow:*

$$DF = \frac{136\,minutes}{17.5\,days \times 480\frac{minutes}{day}} \times 100\% = 1.6\%$$

4. *OBC (including $CT_{effective}$ for workstation 1 and workstation 4 from question 5):*

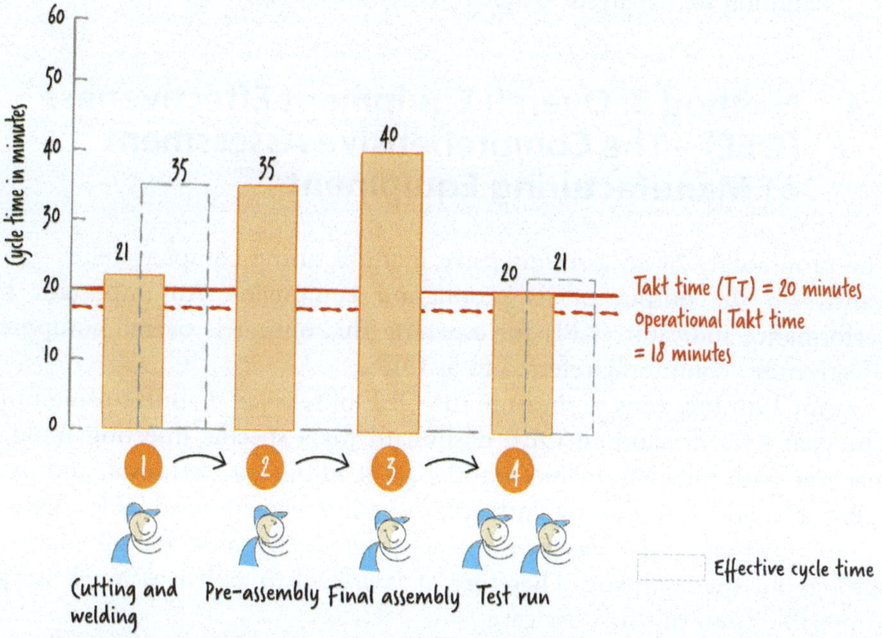


*Short-term measure: doubling staff capacity in stations 2 and 3.*

5. *Effective cycle times:*

$$CT_{effectiv,St.1} = \frac{CT}{availability \times yield} = \frac{21\,minutes}{0.80 \times 0.75} = 35\,minutes$$

$$CT_{effectiv,St.4} = \frac{CT}{availability \times yield} = \frac{20\,minutes}{0.95} = 21\,minutes$$

6. *Improvement options, especially in station 1:*

- *Eliminate waste through reinvestment and by analyzing the root cause of quality issues.*
- *Consider short-term doubling of employee capacity and likely machinery capacity, or in combination with process improvements.*
- *Increase the level of automation in conjunction with a part-time support employee.*
- *Outsourcing of one or several production steps.*
- *A combination of all of the before mentioned measures.*

- Conducting improvement workshops to reduce waste in all four areas using the following methods: general waste analysis, process map, handling step analysis, spaghetti diagram, or A3.

## 2.4 Method 3: Overall Equipment Effectiveness (OEE)—The Comprehensive Assessment of Manufacturing Equipment

The profitability of an asset-intensive manufacturing company relies significantly on the output of its production equipment. An important key performance indicator (KPI) for assessing this output is overall equipment effectiveness, commonly referred to as OEE.

LeanClean Inc. aims to measure the OEE of selected manufacturing units. The goal is to develop an OEE dashboard for a specific injection molding machine with fully automated robotics for loading, parts removal, and packaging. The goal is to transparently identify improvement needs. After gaining initial experience, the dashboard will be gradually used for all machines throughout the company. Therefore, it is crucial to gather as much insight as possible from this first use case.

The table below summarizes the reported data for week 37. This is the sum of 5 working days in a 2-shift operation, in this week.

Data export from LeanClean ERP - all longer interruption times are booked to the order - scanning of the barcode order at start interruption and also scanning of the barcode at stop interruption

| | Documentation of activities / times (in week 37) | Time | Comment |
|---|---|---|---|
| 1 | 2 shifts @ 8 hours | | Plan: 06:00 - 14:45 / 14:45 - 21:45 *) |
| 2 | Two times 15' break time per shift | | 15' Break paid, machine is standing |
| 3 | Per 8 hours, 45' Lunch/Dinner break | | 45' Lunch/Dinner break, stamp out |
| 4 | Weekly planned maintenance | 01.00 | Performed once a week |
| 5 | Tool breakage - 4 times per week | 02.03 | |
| 6 | Setup 5 times - total | 01.56 | Set up times are booked to the order |
| 7 | Allowance times - depending on shift, interruption total | 01.05 | Employee turns off machine |
| 8 | Interruption due to tool adjustment (2 x) - total | 00.32 | |
| 9 | Machine stopped - employee removed | 02.00 | |
| 10 | Yield: 98% of all 17 orders this week | | Was a good production week |
| 11 | Quality department stops machine for inspection | 00.15 | |
| 12 | Quantity actual vs. target over the week: 97% | | Machine speed lower |

Legend: 01.05 means 1 hour and 5 minutes

*) Agreement with the people to take dinner break and one 15' break at the end of the shift, machine runs till 21:45

For the first time, LeanClean Inc. additionally read out machine interruptions from the control system, which turned out to be an elaborate project. Contrary to the claims made by both the machine and controller manufacturer, LeanClean Inc. had to program this interface itself. Fortunately, a software engineering student from OST University could handle this task as part of an internship, successfully programming the interface for the first machine—a commendable achievement by this aspiring software engineer. Based on the above-mentioned exports from the ERP system, the data provided new insights into the many small stops. These so-called "short stops" should obviously not be neglected.

**Reasons for Changes:**

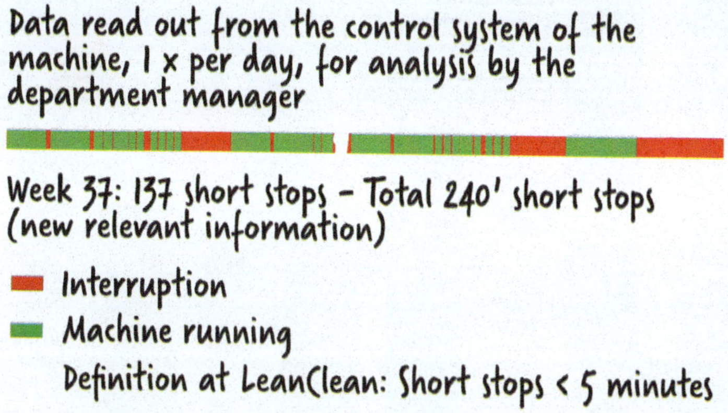


***Task***

1. *Calculate the OEE based on the associated utilization, performance, and quality factors. Consider both the ERP report and the short stops. At LeanClean, a machine's "target runtime" is defined as the length of the shift minus planned downtime, such as employee breaks or scheduled maintenance. Evaluate the OEE and ascertain if it is a good value. What could be done to improve the OEE?*
2. *It is often argued that setup times should be subtracted from the target runtime. While setup truly depends on the product mix, the primary concern is not the absolute OEE value but rather the insights it provides into potential areas for improvement. Nevertheless, how will the OEE change if the target runtime is reduced by the setup times?*

## *Solution*

1. *The total OEE is 78%, the utilization factor is 82%, the performance factor is 97%, and the quality factor is 98%. The OEE is a good value, whereby further optimization of the utilization factor is quite possible. In addition to the optimization of management/planning factors, the reduction of short stops is a possible optimization direction. However, the effort required for improvement is high, as many factors influence the short stops.*

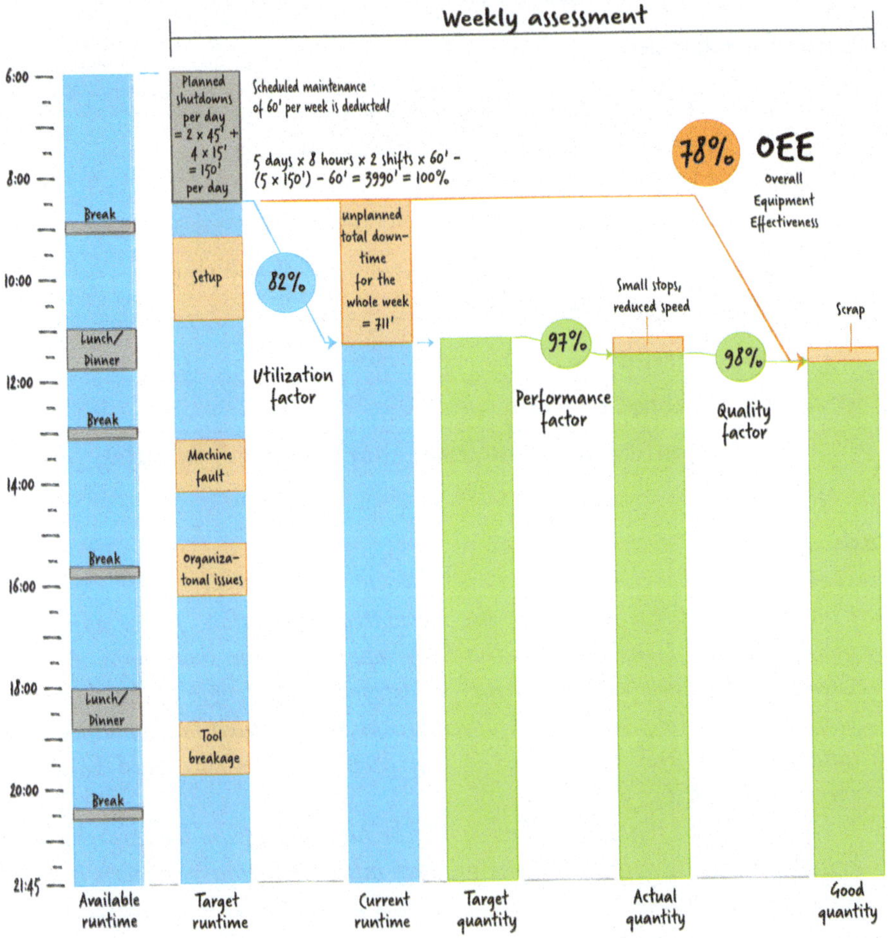

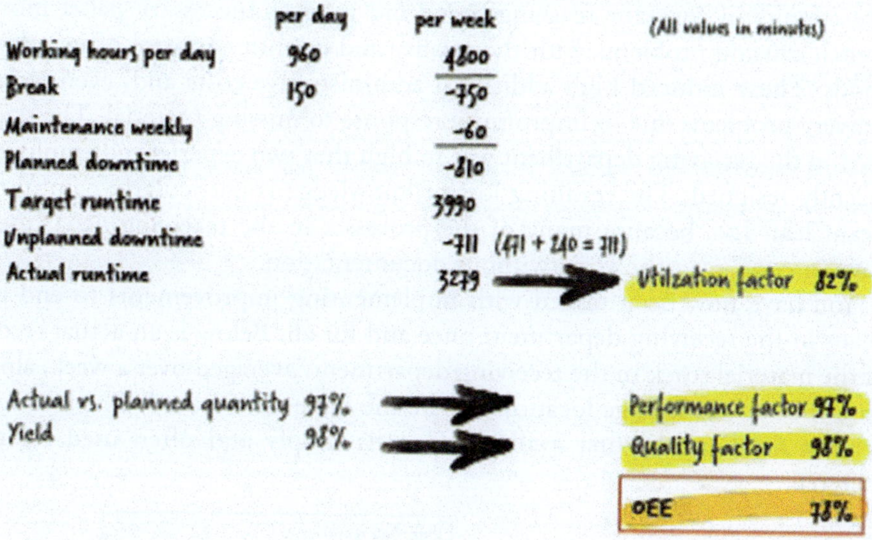


2. *Including setup in planned downtime (instead of unplanned downtime) reduces the target run time by 116 to 3874 min. This increases the utilization factor to 85%, which slightly increases the OEE to 81%.*

## 2.5 Method 4: Handling Step Analysis

Handling step analysis is a methodology that examines the processes supporting production. Each value-added activity in production is associated with many processes, such as transportation, inspection, unpacking, storage, or relocation. The handling step analysis method examines these non-value-added activities. Although these processes do not add value, they are still necessary for the production process. For instance, no production can function without a receiving department or without manual forklifts to deliver products to the production site. However, these processes are always wasteful and should be scrutinized and optimized. Notably, you have to be very vigilant, especially in digitalization age, where many things are "scanned" and "documented". Simplifying these processes could reduce the amount of "scanning" and "documenting" required, leading to fewer errors and a focus on "doing the right thing" instead of "doing the wrong thing efficiently".

Due to constant growth, LeanClean Inc. decided to carefully examine their receiving department. Over the past year, the receiving and the entire logistics process had become a real problem. At times, the receiving area is chaotic, with pallets of material sitting around for days, clogging the space.

This results in constant rearrangement and many unnecessary pallet movements, causing problems with the quality and quantity control of incoming goods. These induced high additional administrative costs and, even worse, delivery problems due to impromptu response to missing material. The stress level in the receiving department was so high that two experienced employees recently resigned. The resulting loss of knowledge is particularly painful for LeanClean Inc. because many of the processes in the receiving department have "grown" over the years without documentation.

You have now been tasked with implementing improvements to end the chaos in the receiving department once and for all. Below is an actual record of the material stock in the receiving department, averaged over a week, along with the various storage locations where the material is stored. The locations are not fixed, as the first available space is simply and often used for the material.

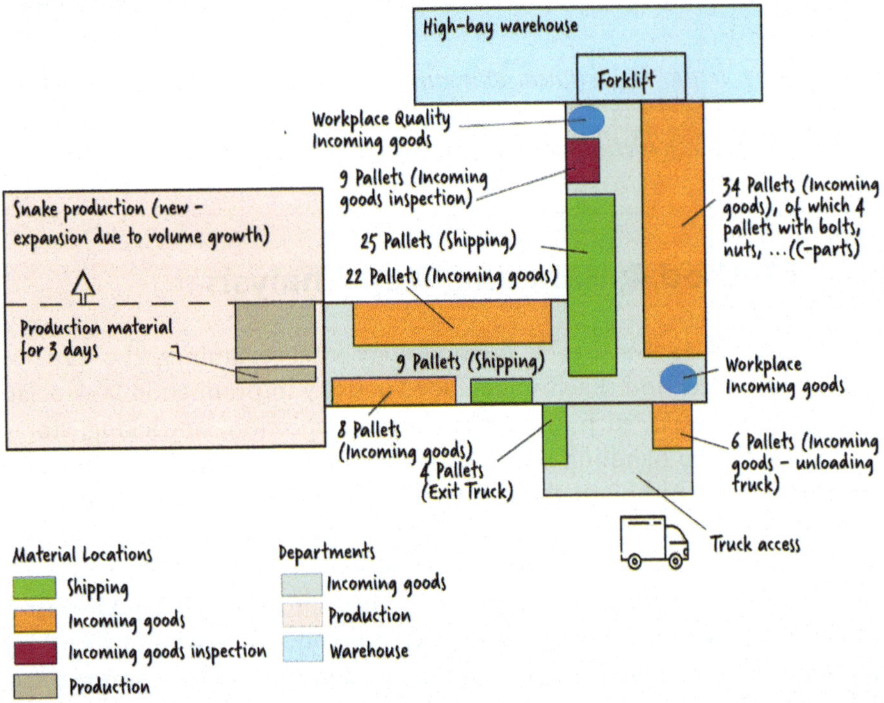


In total, 70 pallets of material are available for storage in the receiving area, another 9 pallets are in the receiving inspection area, and 38 pallets are awaiting shipping and logistics. In addition, discussions with employees have identified significant waste in these areas, as highlighted by the waste flashes. Many handling steps are completely unnecessary—the amount of waste therefore came as a big surprise to management:

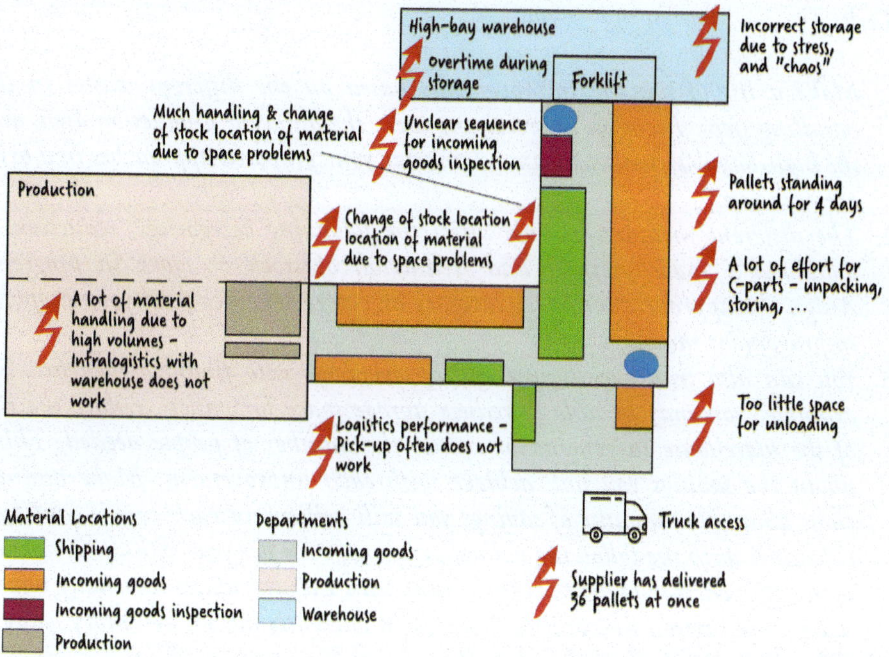


This "as-is" assessment of the warehouse entrance, adjacent production area, and high-bay warehouse serves as a starting point for improvement. Let's get started.

- It is clear that the current situation in receiving and production requires significant enhancement. Optimizing the waste in these handling steps necessitates a comprehensive overhaul of the material flow and logistics, including internal processes and those extending to suppliers. To ensure this optimization is future-oriented, the following data has been identified: The number of pallets delivered to the receiving dock daily currently varies between 35 and 75 pallets; LeanClean Inc. assumes an average of 65 pallets. In the future, you want to dimension the receiving dock to handle a maximum of 50 pallets per day.
- Today, 30 to 50 pallets await shipment in the outbound area, with an average of 45 pallets. In the future, this area may contain a maximum of 35 pallets in the future. The daily logistics processes must be adapted.

### *Task*

1. *Make a list of possible improvements based on the different wastes in the handling steps. Focus on the receiving area, the adjacent production area, and the high-bay warehouse. Please summarize your proposed approaches in a brief presentation.*
2. *The different measures mainly affect the following categories: "reduction of workload", "space saving", and "reduction of stock or work in progress". Assign the appropriate impact category or categories to each of your proposed improvement measures.*
3. *You can also translate the list of improvements into financial benefits. For example, you may be able to reduce storage space for pallets, reduce the size of the warehouse in general, or reduce the number of people needed. Think about the savings you will achieve with each improvement. Make assumptions about the amount of savings you will achieve in each case. LeanClean calculates with the following values: personnel costs per year €60,000, average value of goods per pallet €2000. LeanClean Inc. calculates internally with a warehouse interest rate of 7%. LeanClean currently has an inventory value of €8 million in the high-bay warehouse.*

### *Solution*

1. *Here are 15 potential improvement actions:*
   (a) *Introduction of Kanban supply for production (assembly cells) for the Snake.*
   (b) *Introduction of a milkrun for material flow in assembly.*
   (c) *Deliver material directly to a separate storage location before assembly. The Snake parts are no longer stored in the high-bay warehouse.*
   (d) *Elimination of a general storage location for incoming pallets in the receiving area.*
   (e) *Elimination of the general storage location for outbound pallets in the receiving area.*
   (f) *Increase space for outbound materials by creating space for 10 pallets.*
   (g) *Increase the space for incoming materials to accommodate 10 pallets.*
   (h) *Create and mark space for outbound materials for delivery by establishing a clear delivery sequence for delivery and, if possible, arrange materials in delivery order.*

(i) *Create a dedicated space for pallets storage, with a maximum capacity of 40 pallets. It is important that no more than 40 pallets are stored because problems are immediately visible when pallet's space runs out.*
(j) *Sequence for storage must be clear—the goal must be to have all pallet spaces empty by evening—increase working time to 10 h per day. Clarify and establish back-up arrangements and storage capacity.*
(k) *Deliver C-parts directly to the production area—clarify and agree on C-parts management protocols, including logistics services with special supplier.*
(l) *Introduce a digital interface between the ERP system and high-bay warehouse to reduce typing and scanning, thereby minimizing storage errors.*
(m) *Introduction of a strict FIFO logic in the incoming goods inspection—three important parts could also be switched to "no incoming goods inspection" (supplier now checks the quality and provides the documentation digitally).*
(n) *Daily agreed deliveries (if possible, always in the morning) with a logistics provider with high flexibility—in case of emergency, a second delivery option at 16:00 was agreed.*
(o) *Agreed delivery quantities with suppliers, including timing of delivery quantities per week, day, etc. This coordination is new to LeanClean. The handling steps in the receiving area and the optimization of the flow of goods in the receiving area have not yet been discussed with suppliers. This coordination has often led to positive feedback from suppliers, as they can now optimize and standardize their own logistics.*

*It has been shown that handling steps are closely linked to overall process improvement. Eliminating waste in receiving, warehousing, and internal logistics is essential, as many "hidden" costs are involved. However, implementing numerous comprehensive changes simultaneously would overburden the organization. A step-by-step approach is recommended (refer to the change management exercise at the end of Chap. 3).*

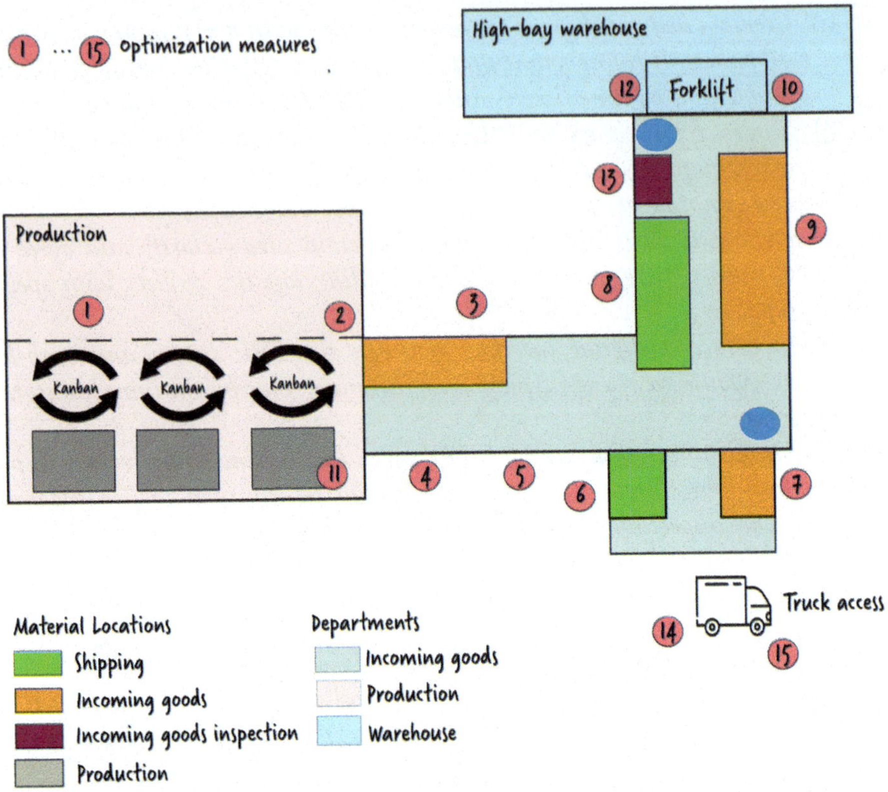


**Tip** Comprehensive organizational and procedural measuresare crucial for reducing the number of handling steps. So, examine it first.

Technical aids, such as scanning solutions and complete handling systems are often discussed and seen as the solution. While these technical solutions can indeed be helpful, they should only be implemented after organizational process optimization.

The milkrun is often seen as an additional and unnecessary handling of material. In practice, it is often argued that this internal logistics causes additional costs and that the "little" material handling can easily be done by the production staff. This work and its impact on the added value of production workers is often underestimated and not measured. It has been shown in many companies that the milkrun results in a large overall improvement in productivity (see also the exercise in Sect. 3.5).

2. *Measures according to impact categories:*

| Measure | Reduction in workload | Space saving | Reduction in inventories & work in progress |
|---|---|---|---|
| (1) Introduction of Kanban supply for production (assembly cells) for the Snake. | X | X | X |
| (2) Introduction of a milk run for material flow in assembly. | | | X |
| (3) Deliver material directly to a separate storage location before assembly. The Snake parts are no longer stored in the high-bay warehouse. | X | | X |
| (4) Elimination of a general storage location for incoming pallets in the receiving area. | | X | |
| (5) Elimination of the general storage location for outbound pallets in the receiving area. | | X | |
| (6) Increase space for outbound materials by creating space for 10 pallets. | | X | |
| (7) Increase the space for incoming materials to accommodate 10 pallets. | | X | |
| (8) Create and mark space for outbound materials for delivery by establishing a clear delivery sequence for delivery and, if possible, arrange materials in delivery order. | | X | |
| (9) Create a dedicated space for pallets storage, with a maximum capacity of 40 pallets. It is important that no more than 40 pallets are stored because problems are immediately visible when pallet's space runs out. | X | X | |
| (10) Sequence for storage must be clear—the goal must be to have all pallet spaces empty by evening—increase working time to 10 hours per day. Clarify and establish back-up arrangements and storage capacity. | X | X | |
| (11) Deliver C-parts directly to the production area—clarify and agree on C-parts management protocols, including logistics services with special supplier. | X | | X |
| (12) Introduce a digital interface between the ERP system and high-bay warehouse to reduce typing and scanning, thereby minimizing storage errors. | X | | X |
| (13) Introduction of a strict FIFO logic in the incoming goods inspection—three important parts could also be switched to "no incoming goods inspection" (supplier now checks the quality and provides the documentation digitally). | X | | X |
| (14) Daily agreed deliveries (if possible, always in the morning) with a logistics provider with high flexibility—in case of emergency, a second delivery option at 16:00 was agreed. | | X | X |
| (15) Agreed delivery quantities with suppliers, including timing of delivery quantities per week, day, etc. | X | X | X |

3. *The financial impact of the improvements:*

*The various measures implemented reduced the workload significantly. Lean-Clean Inc. eliminated one (1) employee in the receiving department (from 4 to 3 employees). The improved organization has induced a significant increase in efficiency, benefiting not just the warehouse employees who handles tasks but also many others who have been relieved of certain tasks. This effort is also included here. As a result, LeanClean Inc. is saving €60,000 per year.*

*The reduction in pallet spaces—15 for incoming material and 10 for outgoing material—amounts to a total decrease of 25 pallet spaces. This results in an annual saving of 25 pallets at €2000 × 7% storage interest = €3500 savings per year.*

*Improvements in logistics and cordination with suppliers and logistics providers for transportation have enabled LeanClean to reduce inventory by €1 million per year. This saves €70,000 per year in inventory costs.*

*The total savings amount to €133,500 (€60,000 + €3,500 + €70,000), which significantly exceeds the investment costs. Finally, the changes primarily required internal resources for design and implementation, with the ERP/warehouse management system costing just under €40,000. All in all, the payback period is less than a year. Wow—projects like this keep LeanClean Inc. moving forward.*

## 2.6 Method 5: Operator Balance Chart (OBC)—Balancing Work Content is Important

As a result of a comprehensive value stream analysis of LeanClean's entire value creation process for the Elephant, you and a team of experts have developed several Operator Balance Charts (OBC) for the various assembly cells. In Japan, an OBC is also known as a "Yamazumi Board". The objective of using an OBC is to optimize the flow of materials and information by strategically scanning the process to effectively identify and eliminate waste.

**OBC of an Assembly Cell**

Let's focus on the first assembly cell of the Elephant to create an OBC. To do this, you visit the shopfloor to meticulously record all work steps, processing times, and cycle times. The assembly cell consists of five sequential workstations, as shown below. The stations are staffed with either one or two operators. The columns represent the cycle times per station. The cycle time CT is the total time elements required to complete the various assembly steps at a workstation—for one Elephant to be assembled by the indicated number of workers.

Most of the predominantly manual operations add value to the product, while others can be considered wasteful on closer inspection. Based on the delivery quantities agreed with the sales department, a customer takt time of 50 minutes is calculated for this assembly cell of the Elephant. In the OBC, this takt time is compared to the cycle times of the successive assembly workstations.

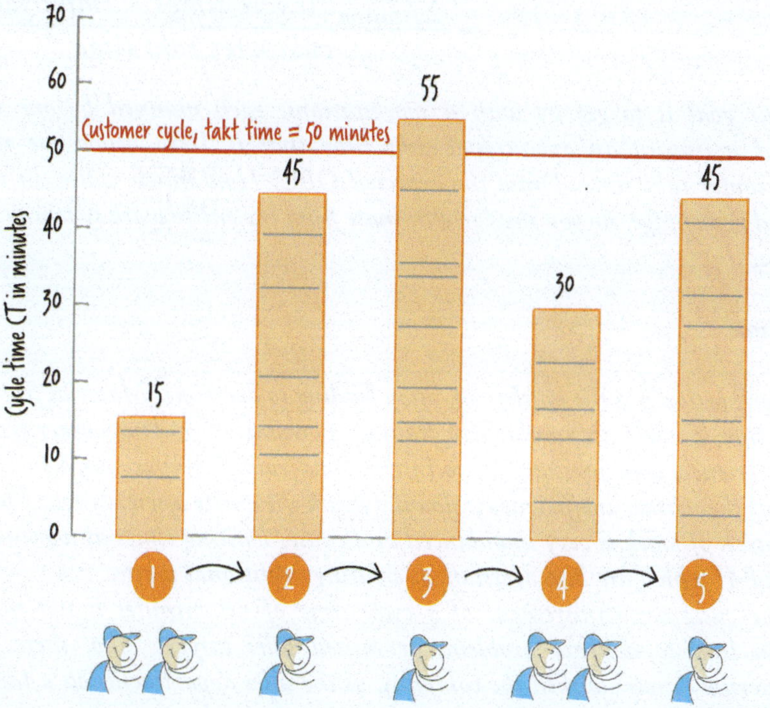


***Task***

1. *Analyze and discuss the situation of this assembly cell for LeanClean's Elephant. First, describe the cycle time differences between the workstations of the assembly cell. Identify waste based on cycle times and number of workers. It is also interesting to discuss how these varying cycle times affect the material flow.*
2. *What are the consequences of workstation 3's cycle time exceeding the takt time?*
3. *Use the following formula to calculate how many assembly workers you would actually need to complete the total workload in an optimal case:*

$$Employees\ required = \frac{\sum process\ times}{customer\ takt\ time} = \frac{\sum_{station\ 1}^{station\ n}(cycle\ times \times employees)}{customer\ takt\ time}$$

4. *Your goal is to get by with 4 workstations, each manned by one worker, and maintain an operational takt time that is 10% below the required customer takt time. Draw the expected OBC. How much waste, in minutes and percentage, do you need to eliminate from the entire work system to achieve this?*

***Solution***

1. *Workstation 3 is the bottleneck, often leading to overtime. Material from workstation 2 likely accumulates in front of workstation 3. The worker frequently experiences time pressure due to instructions from the supervisor, who indicates that the worker inefficiency is slowing down the entire assembly cell. The distribution of work is very unbalanced and causes waiting times at workstations 1 and 4, which are also overstaffed, creating additional waste.*
2. *The customer requirement of a cycle time of 50 minutes is not met. This can lead to delivery problems or missed sales targets. Also, there is often internal "stress" within the company, as the sales department sets a target that production cannot meet.*
3. $Number\ of\ Employees = \frac{(2\times15+45+55+2\times30+45)\,minutes}{50\,minutes} = \frac{235\,minutes}{50\,minutes} = 4.7$
   *The production manager recognizes that the cycle time targeted by the sales department is basically possible: less than 5 employees would be needed, but 7 are used. Assuming an appropriate organization, 2 employees can be saved in this assembly process.*
4. *New operational cycle time = 50 minutes − 10% = 45 minutes as the regular maximum workload per work cycle. New work content would now be 45 minutes for the 4 assembly workstations (or employees) = 180 minutes. Waste of (235 − 180) minutes = 55 minutes or 23.4% must therefore be eliminated from the entire process. This is certainly feasible but requires much work and change.*

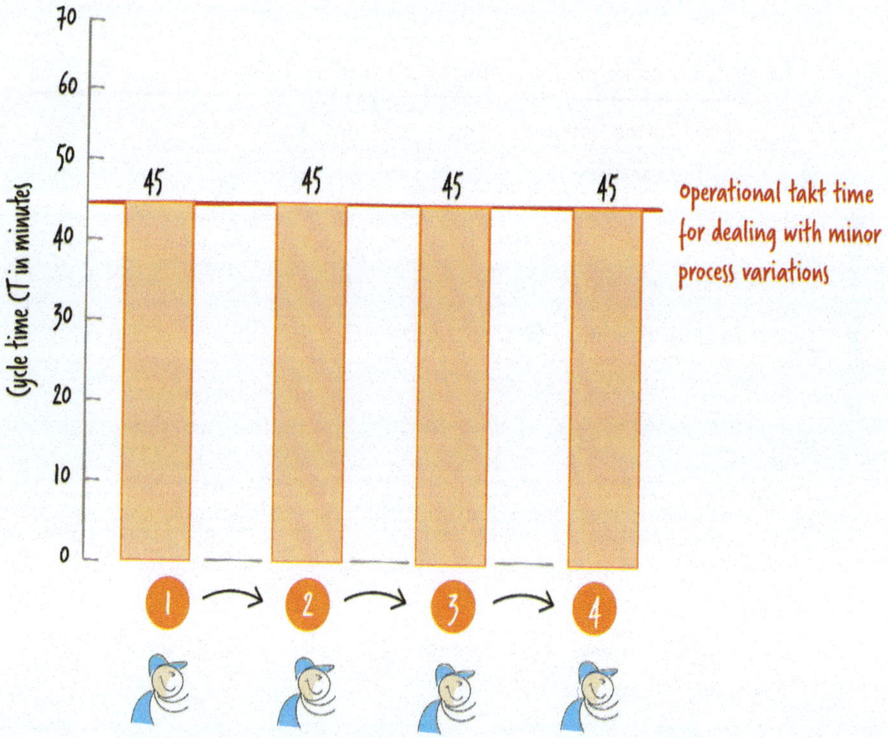


## OBC with a Centralized Workstation

In the second assembly area of the Elephant, the OBC is similar to the previous one, but it has some special features. First, this assembly cell has an integrated process (spot-welding) that is required for all LeanClean Inc. products but is only performed in this cell. Thus, workstation 1 handles spot-welding processes for all value streams in the plant.

In addition, at workstation 2, an employee manages two module sub-variants in parallel but in a time-shifted way due to technical reasons (bonding process). Both sub-variants are manufactured alternately and in equal proportions. The process time consists of 30 min of assembly operation (the cycle time) and a fixed 30-min curing time for the adhesive. For all subsequent process steps, it is irrelevant which Elephant variant is being processed.

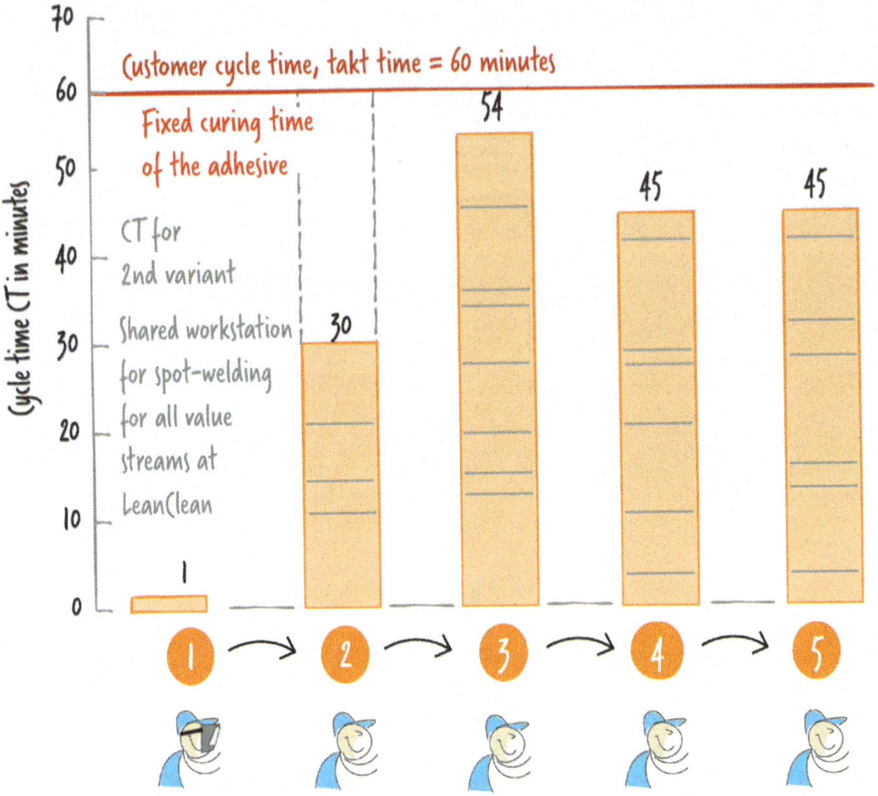


**Task**

1. *Analyze and discuss the situation, especially the cycle time differences between the different workstations. What about other forms of waste and, in particular, employee capacity? How do the peculiarities of stations 1 and 2 affect the rest of the process?*
2. *Calculate how many workers you would actually need in a best-case scenario.*
3. *The goal is to manage the workload with three stations and a staffing capacity of 2.5 employees to meet the operational takt time. Draw the resulting OBC. How much time must be saved at each workstation?*

***Solution***

1. *With 54 min, station 3 has the longest cycle time. Although workstation 2 considers a one-off start-up cycle for curing the adhesive, the cycle time for each module is 30 min thereafter. However, the CT differences between the stations tend to result in increased work in process (WIP) before station 3 and underutilized employees in stations 4 and 5. Integrating spot welding into the value stream is questionable at this point. Consequently, you and your team decide to maintain the current setup for now and establish a Kanban pull process between stations 2 and 1.*
2. *The work content relevant for the production flow amounts to (30 minutes + 54 minutes + 45 minutes + 45 minutes = ) 174 minutes distributed over 4 stations (excluding station 1), so 2.9 employees are required. However, 4 employees are assigned to the work system, i.e., 1.1 employees are too many.*
3. *The target takt time for stable operation is 54 minutes, which would require 3.2 employees (174 minutes ÷ 54 minutes). However, managing with only 2.5 employees requires considerable changes. So, let's closely examine this. Station 3 can remain unchanged since its cycle time of 54 minutes is fine. At station 2, improvements of 6 min are necessary in the operative work process, as the 30-minutes curing time of the adhesive cannot be reduced further for technical reasons. In addition, you can take on additional tasks for other operating areas. Stations 4 and 5 would have to be combined to form a new workstation 4*. However, the cumulative time savings required in workstations 4 and 5 are considerable: 36 minutes or 40% of the current 90 minutes. Achieving these savings will most likely require an investment in automation.*

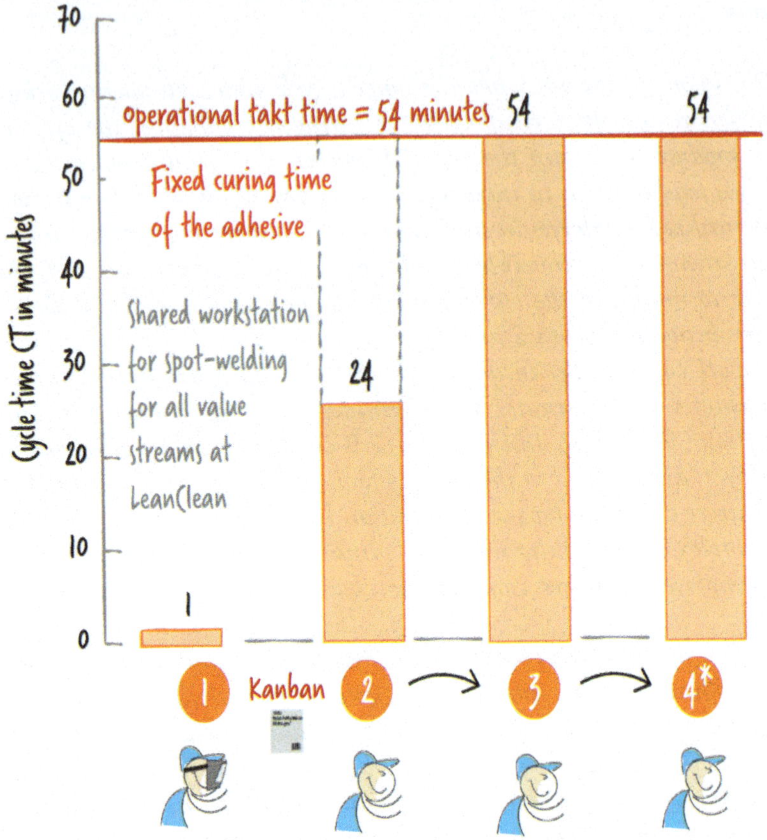


## 2.7 Method 6: Spaghetti Diagram Visualizes Waste by Motion

So far, you have learned that a high proportion of waste is caused by motion, i.e., walking. To meticulously analyze this, we can use the simple but effective "spaghetti diagram" method. Of course, the goal is to avoid paths that resembles spaghetti on a plate. Let's practice this now.

## Walking Distances in an Assembly Cell

While the Elephant is conquering the market for innovative vacuum cleaners, LeanClean Inc. is developing another version of the Octopus. Featuring the BlackHole generator and the groundbreaking MagicMobile technology, this five-armed Octopus only needs to be recharged once a month for the average use in a four-person household. Despite its premium pricing, LeanClean expects this new Octopus variant to have a market potential of 7500 units per year and is testing single-shift final assembly at a single workstation in a dedicated operating area that is connected to the towing train.

The floor plan of the 80 m$^2$ final assembly area, including its main staging areas is shown below. Your task now is to use the assembly work plan to quantify employee walking distances and, ultimately, develop a layout that reduces waste.

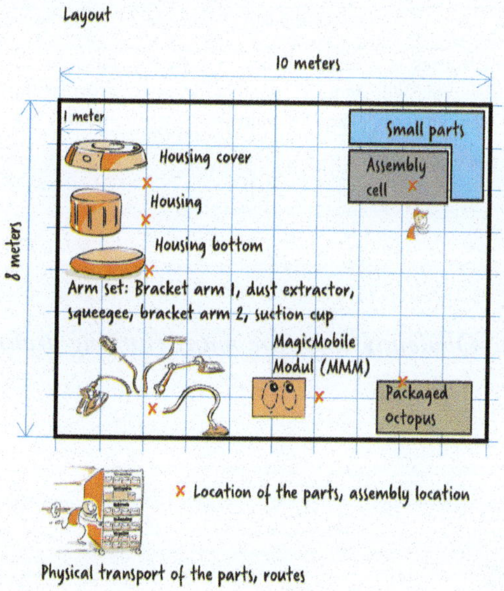


Assembly plan (cycle = 15 minutes)

1. Get housing bottom
2. Prepare MMM assembly
3. Get MMM
4. Assemble MMM
5. Control function
6. Get housing
7. Assemble housing
8. Get housing cover
9. Assemble housing cover
10. Prepare housing for octopus arms
11. Assemble octopus arms (repeat 5 times)
    - Get octopus arm set
    - Pre-assemble arm
    - Assemble arm on Octopus
    - Quality control function
12. Final quality check
13. Dismantle arms and housing cover
14. Pack all components individually
15. Close sales packaging
16. Prepare Octopus for delivery

### *Task*

1. *Visualize the worker's routes with lines in the layout based on the assembly process (the start and end points of each route are marked with "x" signs; we assume that the worker always moves approximately along the straight line). Number each route with the corresponding operation number from the assembly plan. To effectively quantify the routes, use a ruler to measure the*

*distances according to the scale or calculate the hypotenuses of the resulting triangles. Round the measurements to the nearest meter. Determine the total distance $L_{ges}$ per work cycle, per working day (450 minutes), and per year (50 weeks).*

2. *How long does it take an assembly worker to walk the total distance $L_{ges}$ at an average walking speed of 1 meter/second? Determine the cost associated with this type of waste per year based on an hourly rate of 70 €.*
3. *What percentage of the assembly cycle time is walking time?*
4. *Develop a smaller layout, reduced by 3 m in the X-direction and 2 m in the Y-direction, while improving container positioning. How much time can be saved with your optimized layout, or how many more Octopus units can be assembled per day (only if there is customer demand, of course)? What are the additional benefits?*

**Tip** Of course, the best way to record the routes is directly on site, known by Japanese as "gemba". If you do this with the employee, you will acquire many additional information about the process and its weaknesses. These insights provides valuable starting points for further improvements.

## *Solution*

1. *See the summary of results below. Obviously, there are many running around at LeanClean Inc.*

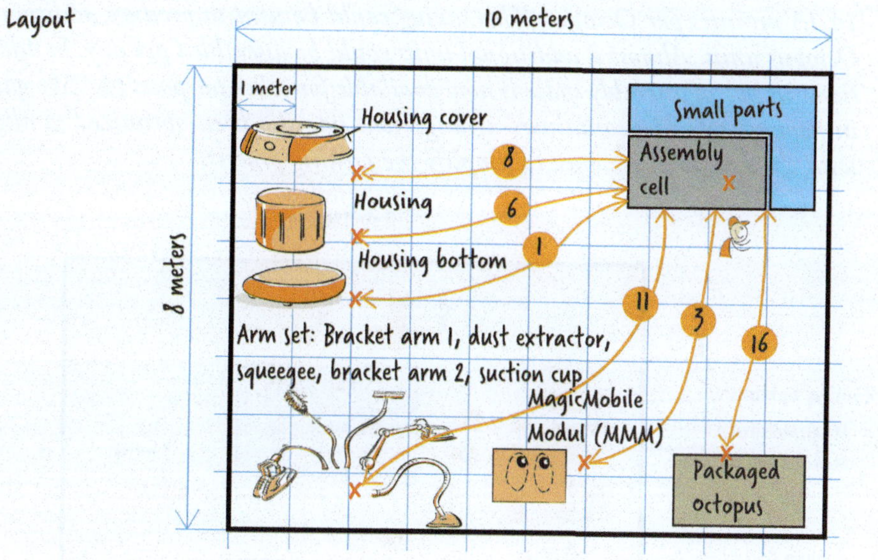


Assembly plan (cycle = 15 minutes)

| | back and forth |
|---|---|
| 1. Get housing bottom | 2 × 7 = 14 meters |
| 3. Get MMM | 2 × 5 = 10 meters |
| 6. Get housing | 2 × 6 = 12 meters |
| 8. Get housing cover | 2 × 6 = 12 meters |
| 11. Assemble octopus arms (total 5 x, per arm) • Get octopus arm set | 5 × 2 × 8 = 80 meters |
| 16. Prepare octopus for delivery | 2 × 5 = 10 meters |

Per octopus assembly cyle ("back and forth") $L_{ges} = 138$ meters

Per day $L_{ges} \times \frac{450 \text{ minutes}}{15 \frac{\text{minutes}}{\text{unit}}} = 4{,}140$ meters (4.14 km)

Per year $L_{ges} \times 30 \frac{\text{units}}{\text{day}} \times 5 \frac{\text{day}}{\text{week}} \times 50 \text{ weeks} = 1{,}035$ kilometers

2. *Time requirement:* $t_{year}$ *= 30 units/day × 138 seconds/unit × 250 days = 17,250 minutes per year or 288 hours per year; cost per year > €20,000*
3. *The walking time proportion is about 15% (138 seconds walking per unit vs. 900 seconds assembly cycle time). The assembly cycle time per unit is 15 minutes or 900 seconds, of which 138 seconds/unit is walking.*
4. *The improved layout reduces the walking distance to approximately 86 meters per assembly cycle, savings of 52 seconds. The improved process now takes only*

*14.13 minutes per Octopus. This saving could be used to produce additional Octopus units. Almost 2 additional units could be assembled per day. In addition, 38 m² of assembly space is now available for other purposes (47.5% gain in floor space). The containers' arrangement has also been optimized to align with the assembly sequence to simplify the picking process.*

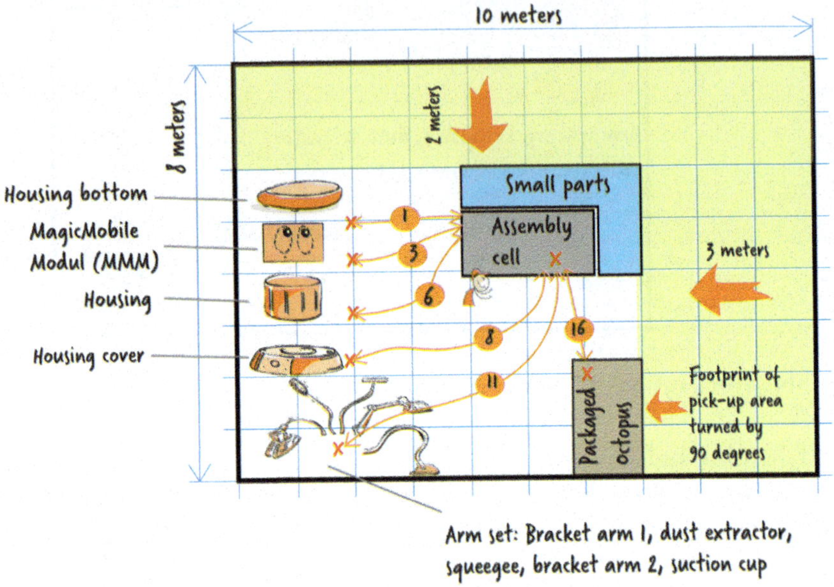


| | | |
|---|---|---|
| 1. Get housing bottom | 2 × 4 = | 8 meters |
| 3. Get MMM | 2 × 4 = | 8 meters |
| 6. Get housing | 2 × 4 = | 8 meters |
| 8. Get assembly bottom | 2 × 4 = | 8 meters |
| 11. Assemble octopus arm (total 5 x, per arm) | | |
| • Get octopus-arm-set | 2 × 5 × 5 = | 50 meters |
| 16. Prepare octopus for delivery | 2 × 2 = | 4 meters |
| Per octopus assembly cycle (for one unit) | $L_{ges}$ = | 86 meters |

➔ Time savings through path reduction (−52 meters per octopus) with new layout results in a productivity improvement of almost 2 additional octopus' per day (28 minutes time savings through reduced walking distances every day) ... and this on 47% less space. What a great result!

## Analysis of the Operation of a Large Machinery

Sometimes you have to deal with operations that are focused solely on keeping a machine running. Rarely is there a work plan for this, but there is a lot of interesting information to be discovered from day-to-day operations. This is the case here.

In order to secure supply, LeanClean Inc. has decided to resume production of the electric motors for the vacuum cleaners ("insourcing"). There are three types of electric motors required for the Snake, Octopus, and Elephant, one of which includes an integrated gearbox. Since the previous supplier wanted to replace its somewhat outdated, fully automated EMM 360 electric motor assembly system with a newer one, the former could be acquired at a good price. This decision was quickly implemented, and the EMM 360 is now located in a former warehouse. Despite its older control technology, the maintenance team successfully commissioned the system.

As depicted in the floor plan below, the EMM 360 is a real "monument" with a total length of 28 m and a width of 4 m. The system is fully automated, and through its plexiglass enclosure, one can see extensive automation and conveyor technology, along with numerous sensors that constantly monitor the availability of parts at the feed points, the production flow, and the product quality. This leads to the problem that got you involved: parts frequently jam at the step conveyors and vibrators, which supply raw materials and individual parts for the production process inside the machinery. The grippers for the motor armatures, motor housings, and gear covers also struggle to handle the components in the trays. And unfortunately, the machinery is not yet fully connected to the logistics train, necessitating the manual supply of materials and removal of finished motors via a central logistics area (input/output buffer).

To deliver the required quantities, the EMM 360 operates in two shifts, from 6:00 a.m. to 10:00 p.m., totalling 16 h per day. Each shift has two employees who manage the system's operation, including parts supply at feed points, container changes, troubleshooting, and quality issues.

In collaboration with two employees, they agree that the entire process needs improvement. With appropriate integration of the logistics train, one employee can operate the system per shift. The key to identifying waste is to track walking distances. To get a representative understanding of the activities in the area, you accompany each employee on each shift twice for 20 min, in a total of eight hand-sketches. The spaghetti diagram below shows a typical 20-min process, which will serve as the basis for the following calculations.

*Task*

1. *First, estimate the total distance walked by a single employee during a 450-min shift. Then, estimate the total distance $L_{ges}$ walked per day by all employees. To do this, trace the marked walking routes from beginning to end with a pencil, counting the number of 1-m squares crossed; partial squares count as full squares, which approximately compensates for the longer diagonal distances. Again, assume an average walking speed of 1 m/s.*
2. *Calculate the amount of daily working time spent walking and determine what percentage of the total working time this represents. Is one employee per shift sufficient as you had hoped?*
3. *What other disadvantages are associated with the close involvement of employees in maintaining EMM 360's operation? What next steps do you suggest?*

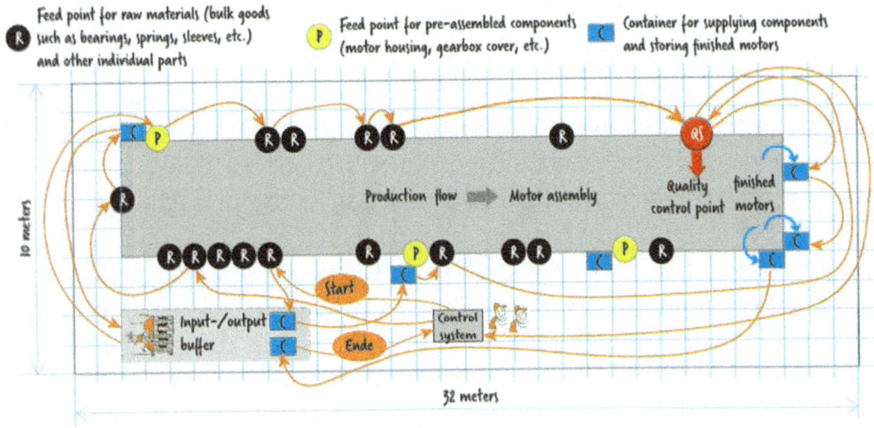


**Tip** The example given is only a snapshot and is not a strict time study according to REFA standards for use in a work plan, nor a methodical, formal activity-sampling study. The latter would have to extend over many weeks for statistical validation.

The classic way for recording walking routes on-site using a floor plan of the area on a clipboard and to enter the employee's routes directly into the floor plan by walking alongside them. It is also advisable to track the walking routes simultaneously using a smartwatch and/or a suitable smartphone app. A GPS signal combined with a mapping app can also be very useful, especially in open spaces.

***Solution***

1. *Total distances according to the sphaghetti diagram:*

$$L_{ges} = 175\,meters \times \frac{450\,minutes}{20\,minutes}$$
$$\approx 3900\,meters\ per\ employee\ and\ shift$$

$$L_{day} = 2\,employees \times 2\,shifts \times 3900\,meters = 15,600\,meters$$

2. *Time required at 1 meter/second:* $t_{day} = 15{,}600\ s = 260\ min = 4.3\ h$ *per day*

   *This corresponds to about ½ employee per day. In the current situation, even if walking distances were completely eliminated (unrealistic), ensuring an operation with only one employee per shift remains impossible. Further improvements are necessary.*

   *The percentage of total daily working hours is:* $\frac{260\,minutes}{4 \times 450\,minutes} = 14.4\%$
3. *Other drawbacks: Employees have limited time and inclination to identify the root causes of short downtimes and quality problems. Next steps: Introduction of an OEE for the system; fully integrate all feed points and container locations with the traffic route, as well as simultaneously adjusting the respective supply volume of the feed points to the route cycle. In addition, consider technical improvements to the system.*

## 2.8 Method 7: Pareto Chart—The 80/20-Rule

Implementing Lean requires a strong focus on the essentials. The Pareto chart is a central Lean method to achieve this, which is also widely used in quality management. Again, the goal is to focus on the most important defects, with Lean's zero-defect principle exemplifying this as a clear standard: every defect is waste. This leads us to the Pareto principle.

The Pareto principle states that 20% of errors or problems cause 80% of waste. Let's practice the application of this principle now. The Pareto chart often serves as an initial analysis tool for applying more advanced quality management methods.

### The Pareto Chart

The following chart shows a Pareto analysis by LeanClean Inc. for an important plastic part:

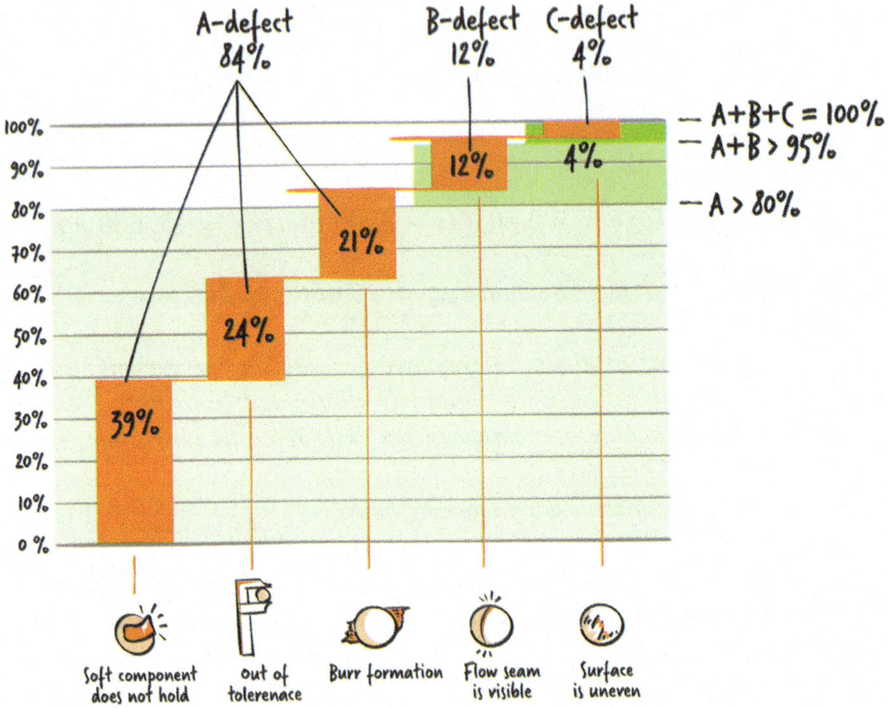


As shown, the three defect types—"soft component does not hold", "out of tolerance", and "burr formation"—account for approximately 80% of the defects. These three types of defects are the A defects, known as the most important defects, and therefore the focus must be on these three issues.

> **Tip** The Pareto chart is not an exact science and does not say that the A-errors have to be exactly 80%. Instead, you decide for yourself what the A-errors are, and then the total is "roughly 80%". Certainly, the A-errors should not exceed 90%.
>
> While the Pareto chart is often employed in quality management, its applications is very useful for other production problems. For example, it is also valuable in inventory analysis as demonstrated in Sect. 2.9.
>
> Modern IT tools make creating a Pareto chart easy. For this purpose, the chart often used is the "inverted waterfall".

## *Task*

1. *Now create a Pareto chart for the defect types in the incoming inspection for the last month. You have analyzed the defect types of the different parts in the receiving department. The result of your analyses is as follows:*

| | Quality cost / Scrap in EUR | In % |
|---|---|---|
| Hole in base plate missing | 21,200 | 18% |
| Surface of brass part scratched | 16,000 | 14% |
| Compression spring lenght out of tolerance | 4500 | 4% |
| Thickness of rubber gasket too small | 780 | 1% |
| Packaging material delivered damp | 1670 | 1% |
| Function test of robot circuit board not ok | 4350 | 4% |
| Color for red Elephant handle not to RAL specification | 9870 | 9% |
| Weight of base plate out of tolerance | 56,340 | 49% |
| Total | 114,710 | 100% |

## *Solution*

1. *The Pareto chart shows where to focus your quality improvement efforts. The three defect types—"weight of base plate out of tolerance", "hole in base plate missing", and "surface of brass parts scratched"—are the three biggest wastes, accounting for 81% of all defects. These are priority areas for improvement.*

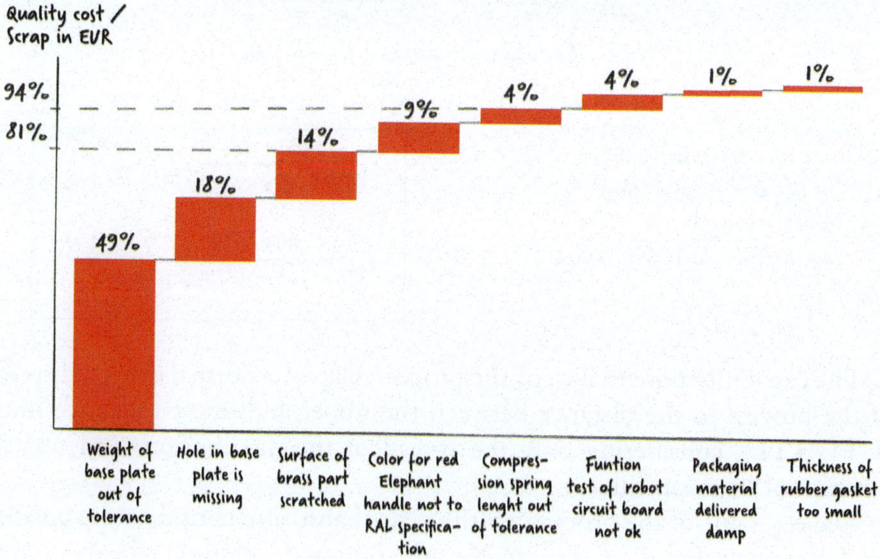

You need to address these A errors, starting with a detailed analysis base plate's weight error.

### Evaluate process capability using $c_p$ und $c_{pk}$

Action was taken based on the Pareto chart, and you've received feedback from the assembly department indicating that defects in this component were detected too late during the assembly process, leading to significant additional costs (including rework and scrap) and delays in customer deliveries. To prevent this waste, you need to gain a better understanding of the problem and the frequency of the defect. For this, you will use the process capability indicators, $c_p$ and $c_{pk}$. Using this understanding of the magnitude's problem, you can engage into a fact-based dialog with your supplier. But what are $c_p$ and $c_{pk}$ anyway?

The $c_p$- and $c_{pk}$ metrics are measures of process capability, showing how well a production process meets specified requirements, such as dimensional accuracy. The data is based on samples.

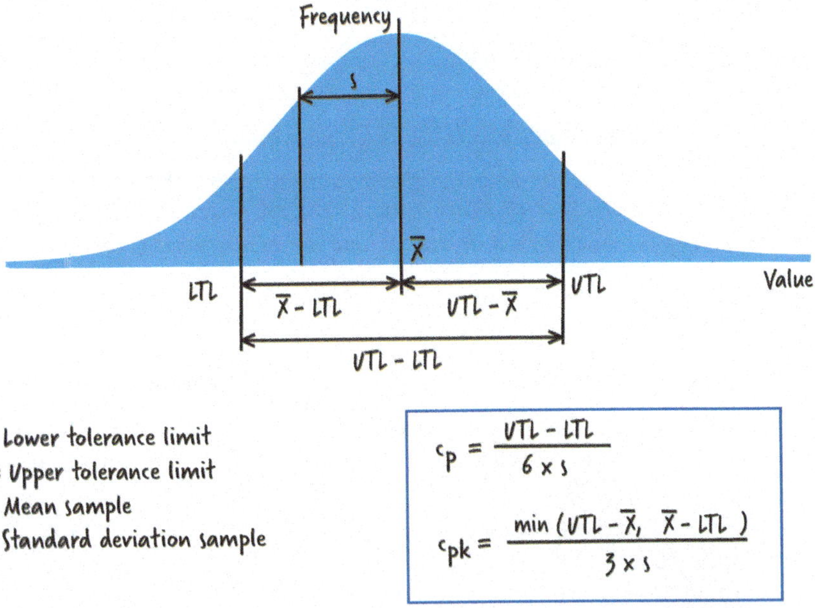


The capability potential $c_p$ of the process relates its normal expected spread of the process to the distance between the upper and lower tolerance limits (UTL, LTL), considering only the spread of the distribution, and not the position of its mean value.

The $c_{pk}$ value of the process, on the other hand, also considers the position of the mean value $\overline{x}$ of the sample and thus its centering. Since $c_{pk}$ only includes the shorter distance to the mean in its calculation, the $c_{pk}$ value will always be smaller than the $c_p$ value for processes that are not perfectly centered.

In principle, a process is considered capable if both the capability that indices $c_p$ and $c_{pk}$ are greater than 1. In many cases, the stricter specification of 1.33 is used as the criterion for process capability, which corresponds to 99.997% of non-defective items.

You analyze the incoming inspection data for all parts over the last 3 months and weigh all the parts in the warehouse. Also, summarize the following facts about the individual weighed parts. You then obtain the mean $\overline{x}$ and the standard deviation $\sigma$ for both suppliers A and B over all measuring points. Supplier A has been an existing supplier for years while supplier B is new, because a "two-supplier strategy" has recently been implemented.

### Task

*First the facts: The weight of the base plate is 800 g—this is the specification for the weight of the base plate. The design department has set the tolerance for the base plate at ± 50 g based on quality requirements (stiffness) and the maximum acceptable weight for a portable vacuum cleaner.*

*You have now determined the following data from a random sample for both suppliers:*

- *Existing supplier A:* $\overline{x}A = 801.25s\ A = 10.71$
- *New supplier B:* $\overline{x}B = 813.30s\ B = 17.80.$

1. *Determine the UTL (Upper Tolerance Limit) and LTL (Lower Tolerance Limit) for the base plate.*
2. *Calculate the short-term process capabilities for supplier A and supplier B. Which supplier do you prefer and why?*

### Solution

1. *The specification for the base plate is that the weight must not deviate more than 50 g up or down from the target value.*

   *Therefore, the specification limits are as follows:*

$$UTL\,(Upper\ Tolerance\ Limit) = 800g + 50g = 850g$$

$$LTL\,(Lowert\,Tolerance\,Limit) = 800g - 50g = 750g$$

2. *For the existing supplier A, the $c_p$ und $c_{pk}$ are as follows:*

$$c_pA = \frac{850g - 750g}{6 \times 10.71g} = 1.56$$

$$c_{pk}A = \frac{\min(850g - 801.25g\;or\;801.25g - 750g)}{3 \times 10.71g} = \frac{78.75g}{32.12g} = 1.52$$

*For the new supplier B, the $c_p$ und $c_{pk}$ are as follows:*

$$c_pB = \frac{850g - 750g}{6 \times 17.80g} = 0.94$$

$$c_{pk}B = \frac{\min(850g - 813.30g\;or\;813.30g - 750g)}{3 \times 17.80g} = \frac{36.70g}{53.40g} = 0.69$$

*Both indicators are significantly better for supplier A: smaller spread ($c_p$) and significantly better centering ($c_{pk}$). If supplier B wants to establish itself as a second supplier in the long run, it must work on its process stability.*

## First Pass Yield and Yield Calculation

The Pareto chart identified "missing hole in base plate" as the second most critical challenge. The assembly manager and the supplier agreed that Lean-Clean would drill the hole during the assembly process and charge the supplier for the extra work. While this is the best solution for both parties, the assembly manager wants to precisely monitor this additional process. He intends to measure both the First Pass Yield at the receiving dock and the yield after redrilling. But how are these KPI determined?

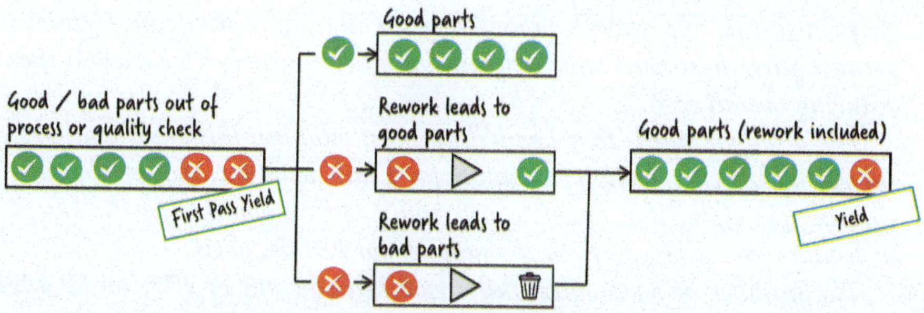


First pass yield is the proportion of good parts in the first process (without rework), in relation to all parts

$$\text{First Pass Yield} = \frac{4}{6} = 66\%$$

Yield is the proportion of good parts (including all parts that have become „good parts" through the rework process) in relation to all parts

$$\text{Yield} = \frac{5}{6} = 83.3\%$$

The First Pass Yield is the "harder" measure because it excludes good parts from rework and immediately reveals waste.

Over the last 3 months, 137 parts out of 2890 delivered parts were rejected during the incoming inspection. As agreed, the assembly manager established a rework process, salvaging 54 parts by redrilling the hole.

***Task***

1. *Calculate the first pass yield and the yield for the "hole" feature of the base plate.*
2. *LeanClean Inc. has implemented this rework process temporarily. Would you continue to maintain this additional process? What are the pros and cons?*

***Solution***

1. *The First Pass Yield and the yield are calculated as follows:*

$$First\ Pass\ Yield = \frac{2753}{2890} = 95.3\%$$

$$Yield = \frac{2807}{2890} = 97.1\%$$

2. *At first glance, the rework process benefits are lower than the logistics cost because parts don't need to be returned back to the supplier for rework, thereby reducing overall costs.*

   *The disadvantage is that LeanClean must establish this additional process, increasing the complexity of assembly for LeanClean Inc, despite the supplier covering the costs. Also, LeanClean is now partly responsible for the baseplate, as it takes over a process from the supplier for certain parts.*

   *The number of parts successfully reworked (54 out of 137, or 39.4%) is relatively low and LeanClean is neither equipped nor experienced with this process. Therefore, it's worth questioning whether the effort and increased complexity justify this process that initially appears to have a low total cost—probably not.*

## 2.9 Method 8: Inventory Analysis

Like many companies, LeanClean Inc. faces challenges with high inventory levels. While several measures, particularly in purchasing, have led to some improvements, addressing the root cause requires a deeper look at the processes. Here's how one of the warehouse culprits was approached. Recently, the Octopus has become a bestseller, but one crucial assembly of 20 parts has created significant delivery problems. It's evident that these issues stem from planning inefficiencies, which in turn have contributed to inflated inventory. By analyzing this inventory, you hope to gain important insights. To begin, you download relevant data from the ERP system.

| No | Art.-No. | Description | Yearly consumption in EUR | Stock removal per year |
|---|---|---|---|---|
| 1 | 103456 | Reel | 12,340 | 12 |
| 2 | 104234 | Ball bearing 12 | 1010 | 67 |
| 3 | 102334 | Flange 6 | 45,020 | 12 |
| 4 | 100001 | Housing round | 83,900 | 26 |
| 5 | 103450 | Shaft small | 18,100 | 8 |
| 6 | 101290 | Shaft large | 22,020 | 7 |
| 7 | 105981 | Flange plastics | 3100 | 12 |
| 8 | 102987 | Cover screwed on | 1100 | 12 |
| 9 | 107234 | Cover metal | 8800 | 32 |
| 10 | 106789 | Screw 6 x 30 | 500 | 20 |
| 11 | 100345 | Screw 4 x 25 | 600 | 90 |
| 12 | 100781 | Washer 6 | 400 | 23 |
| 13 | 100894 | Washer 4 | 390 | 87 |
| 14 | 101234 | Rubber seal | 1000 | 10 |
| 15 | 102390 | Gasket | 1190 | 23 |
| 16 | 106900 | Wheel | 23,800 | 4 |
| 17 | 107345 | Nut 6 | 980 | 34 |
| 18 | 100034 | Nut 8 | 230 | 78 |
| 19 | 104561 | Housing cover special | 98,000 | 5 |
| 20 | 104788 | Underbody aluminium | 124,000 | 56 |
| | | Total | 446,480 | |

Stock removals are defined as withdrawals from the warehouse with a defined lot size. The frequency of these withdrawals is an indicator of how often and consistently the parts are requested by the various production areas. At first, you are surprised to discover that these 20 parts account for a stock value of €446,480. A closer look reveals that the withdrawal frequency from the warehouse fluctuates greatly, and is influenced by factors such as parts size, cost, the replenishment time, and batch sizes in both assembly and production. To address this, you decide to categorize the parts into different groups.

***Task***

1. *Make an ABC classification and analysis of all parts, where the most important parts, the A-parts, account for approximately 80% of the stock value. The B-parts make up for around 15% of the stock value, while the C-parts—typically bulk items like screws, nuts, washers, DIN, and other standard*

*components—approximately constitute the remaining 5% of the stock value. This ABC logic is very similar to the Pareto principle.*

2. *You have also recorded the annual stock removal processes as an indicator of how often and consistent parts are requested by various production areas. Based on this, ypu categorise parts as X, Y, and Z, according to the number of withdrawals from stock. You define this as follows: Parts with up to 12 stock removals per year are Z-parts. Parts with 13 to 52 stock removals are Y-parts. Parts with more than 52 stock removals per year are X-parts for you. LeanClean removes X parts from inventory each week.*
3. *Discuss how to plan and control the different parts? Where should the focus be? What can be done to improve warehouse management?*

## Solution

1. and 2. *The following table categorizes the parts in ABC-XYZ:*

| No | Art.-No. | Description | Yearly consumption in EUR | Yearly consumption in % from total | ABC classification | Stock removel per year | XYZ classification |
|---|---|---|---|---|---|---|---|
| 20 | 104788 | Underbody aluminium | 124,000 | 27.8% | A | 56 | X |
| 19 | 104561 | Housing cover special | 98,000 | 21.9% | A | 5 | Z |
| 4 | 100001 | Housing round | 83,900 | 18.8% | A | 26 | Y |
| 3 | 102334 | Flange 6 | 45,020 | 10.1% | A | 12 | Z |
| 16 | 106900 | Wheel | 23,800 | 5.3% | B | 4 | Z |
| 6 | 101290 | Shaft large | 22,020 | 4.9% | B | 7 | Z |
| 5 | 103450 | Shaft small | 18,100 | 4.1% | B | 8 | Z |
| 1 | 103456 | Reel | 12.340 | 2.8% | C | 12 | Z |
| 9 | 107234 | Cover metal | 8800 | 2.0% | C | 32 | Y |
| 7 | 105981 | Flange plastics | 3100 | 0.7% | C | 12 | Z |
| 15 | 102390 | Gasket | 1190 | 0.3% | C | 23 | Y |
| 8 | 102987 | Cover screwed on | 1100 | 0.2% | C | 12 | Z |
| 2 | 104234 | Ball bearing 12 | 1010 | 0.2% | C | 67 | X |
| 14 | 101234 | Rubber seal | 1000 | 0.2% | C | 10 | Z |
| 17 | 107345 | Nut 6 | 980 | 0.2% | C | 34 | Y |
| 11 | 100345 | Screw 4 x 25 | 600 | 0.1% | C | 90 | X |
| 10 | 106789 | Screw 6 x 30 | 500 | 0.1% | C | 20 | Y |
| 12 | 100781 | Washer 6 | 400 | 0.1% | C | 23 | Y |
| 13 | 100894 | Washer 4 | 390 | 0.1% | C | 87 | X |
| 18 | 100034 | Nut 8 | 230 | 0.1% | C | 78 | X |
| | | Total | 446,480 | | | | |

| Yearly consumption -> ABC | | |
|---|---|---|
| > 80% | A | Cost relevant |
| 5% - 80% | B | Medium |
| < 5% | C | C-Parts |

| Removal per year -> XYZ | | |
|---|---|---|
| 1 - 12 | Z | Slow Mover |
| 13 - 52 | Y | Medium |
| > 52 | X | High Runner |

3. *Depending on the category, you use different logistics processes. Parts with high demand or frequent withdrawals are best managed using Kanban–either within production or, ideally, directly from the supplier. You define the Kanban*

*buffers and agree on fixed delivery schedules with the supplier, which usually and significantly reduces inventory levels. Discussions with the supplier allows you to explore further process optimizations, such as packaging and ordering processes.*

*Low-cost parts (C-parts) are managed in a special way. Based on Kanban, C-part containers are filled by weight or simply by the quantity indicated on the container label, eliminating the need to count every single screw. Some suppliers replenish the containers for you according to the Kanban logic. You no longer manage these parts in the warehouse, but have them delivered directly to the production area.*

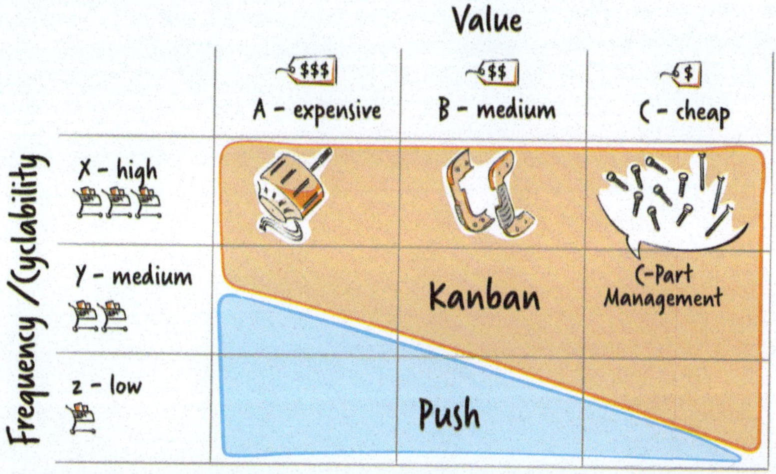


*You should closely examine the so-called "push parts". These parts are best managed in your ERP system. You need to define an order stock for them and ensure that master data such as replenishment times, order quantities and other relevant parameters are reviewed and agreed upon with your suppliers. There is often significant potential for improvement in these areas. For more details, refer to Sect. 3.7, where various push approaches are discussed.*

# 3

# Exercises on Methods for Implementing Principles and Change

## Contents



© The Author(s), under exclusive license to Springer-Verlag GmbH, DE, part of Springer Nature 2025

R. Hänggi et al., *LEAN Production Training for Practice and Study—35 Exercises with Solutions*, https://doi.org/10.1007/978-3-662-70468-4_3

## 3.1 Method 9: 5S—The Basic Method of Lean

The 5S method is about "sorting (out)", "setting in order", "making shine", "standardizing", and "sustaining (the standard)". It is an important Lean method used in any Lean transformation.

**Tip** "Sorting out" is always the first step—removing unnecessary clutter. However, this only refers to clutter that is truly unneeded. The simple solution is always to "throw it away", which is easy, but it's only half the battle. You will find items (e.g., tools, materials, boxes, and packaging) that are not needed daily but are still worth keeping. These must also be sorted out and then organized appropriately.

"Setting things in order" is a central element of 5S. Labeling materials, tools, marking storage locations, or even creating route markings, and escape routes can significantly improve processes by reducing searching times. There are countless ways of creatively "setting things in order"—use colors, signs, symbols, or text to organize your workspace. You will learn a lot about the processes.

With "sorting out" and "setting in order" you have taken the first steps of 5S, but you're not done yet. If there is also dirt or disorder, the next step is "making things shine"—cleaning and tidying up.

"Standardization" is a well-known approach. When you standardize something, you establish a clear baseline that allows the process to be repeated consistently without much fuss. It can then be performed reliably by any employee. Training is also a type of standardization—endeavor to document these training sessions to identify areas for improvement or employees who may need further training. Let's also keep in mind, "once trained" doesn't mean "trained forever". Stick to a regular training routine!

To ensure that your area does not slip back to the old state, the final element is "sustaining the standard". A periodic 5S audit will often help to maintain focus on the 5S. Constant transparency about the current status is important. Always conduct these audits openly and together with the employees.

As part of 5S, you often use the "waste walk" method. As the name suggests, the idea is to "walk around" and look for "waste". It is based on the Japanese idea of "Go to Gemba", meaning "Go to where it happens".

Before conducting a waste walk in a production facility, it's a good idea to practice the method at home. This provides an ideal environment for a waste walk to become familiar with 5S methodology while also identifying and potentially implementing many improvements right away.

The "Waste Walk" is a great tool for identifying waste. Become a waste reduction professional in your home. But before you start, tell your fellow housemates about your plan. This same principle applies when conducting waste walks at work—openness and transparency in this process will lead to success.

Always take photos of the problems during the waste walks. Then, assign each photo a type of waste and suggest a possible solution:

| Photo | Waste | Solution |
|---|---|---|
|  | <br>overproduction<br>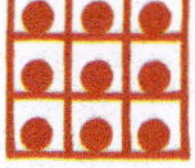<br>Stock | New shelf<br>Clean up<br>Stock reduction - reduction of further shopping ("Eat what is available") |

The combination of different waste symbols and photos helps document the facts, making it easier to communicate the waste ("See, that's what I mean") and the first possible solutions.

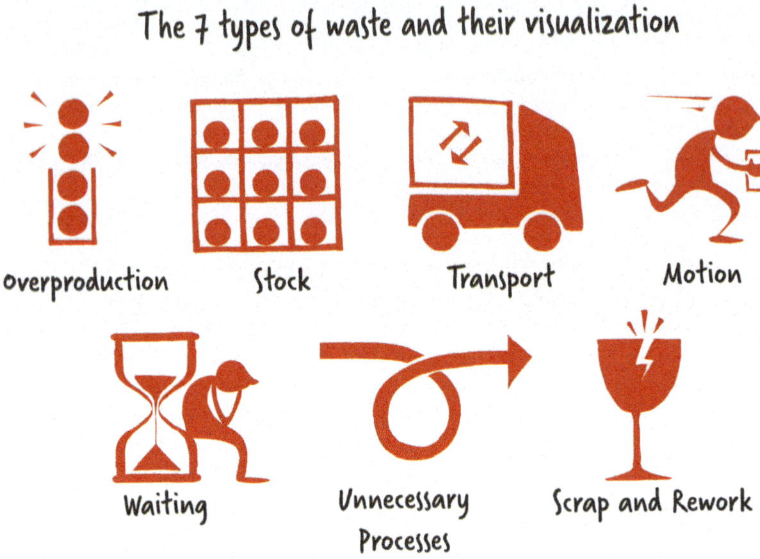


You will see that this simple approach reveals numerous improvement opportunities.

> **Tip** Clutter, or excess material, is often the "inventory" type of waste. However, clutter can also indicate a bad process, e.g., if you have a big mess in your pantry, you might buy jam that you actually still have. Thus, the causal waste type of this clutter is "overproduction" because you are buying or producing jam when you still have jam in the cellar. The solution to this "overproduction" waste would then be, for example, to set up new storage racks and then clean up the storage cellar. This can improve the transparency of the necessary stock. However, this is only one way to improve the situation; there are certainly many other solutions, but It always depends on the specific case.

*Task*

1. *Do a waste walk in your home. Use a table like the one shown above to identify waste and record initial ideas for improvement. Three pieces of information are important:*
   - *Photo*
   - *Type of waste*
   - *Solution.*
2. *You will often find things in the household that have been lying around for a long time (waste type "inventory"). Mark these items with a Post-it. If you use a red Post-it, this is often referred to as "red tagging". You identify things as superfluous although you are not yet sure what to do with them. You have the options of "throw it away", "keep it", "sell it", or maybe even "recycle it". Experience shows that making these decisions as a group leads to better outcomes.*

   *So, why don't you do this red tagging process in your home right now? Take a red Post-it and go ahead. Document your experiences.*

*Solution*

1. *This exercise at home is no different than what you would do in production at your company. You will find many surprising wastes and optimizations. The results of the waste walk should always be discussed openly within the team. Different perspectives make problem solving more robust, and above all, the discussion sensitizes the team - also an important step towards improvement.*
2. *Have you found something surprising that you no longer need and that keeps getting in the way? In the red-tagging process, it is very important that decisions about what to do are made together as a team. Nevertheless, it is especially important to be consistent when it comes to throwing things away. It is often more difficult to discard than to keep.*

## 3.2 Method 10: Zoning—The Optimal Layout Can Be Calculated

Zoning is a method for reducing distances or movements. In this approach, the workplace is organized so that frequently used parts or tools are positioned closely together, while less frequently needed items can be placed further away. This principle can also be used to organize warehouses or production areas. Other methods that support zoning are the spaghetti diagram or the lean shelf. The principle of zoning is also applicable in offices or administrative processes, where document storage should ideally be organized according to the zoning approach.

The most common way to reduce distances on the shopfloor is to create manufacturing cells. In cell manufacturing, all the processes for making a product are in proximity. This creates flow, reduces distances, and improves communication within the cell, leading to a lively exchange of improvement ideas—such as how to further optimize production flow.

The question now is how to group the machines or workstations for different products into cells. We will use the following technique to address this: once you've calculated the row values, rearrange the rows according to these values. This step helps to rank the processes.

**Tip** You start by creating a matrix ("Starting position"), where the processes (e.g., machines, assembly workstations) are listed on the y-axis, and the products are listed on the x-axis.

In the first step, calculate the row value for each row ("1. Calculate row values") by multiplying each row cell by its assigned binary weight, and then, adding the products This is known as binary transformation. For example, in row 1($1 \times 8 + 0 \times 4 + 2 \times 2 + 1 \times 1 = 11$). When assigning binary weights, (for example, $8 = 2^4 = 2 \times 2 \times 2 \times 2$), enter them in descending order above the columns.

The different row values then give you the ranking of the processes ("2. Rearrange rows").

Then, calculate the column values after the binary transformation ("3. Calculate column values").

In the 4th step, you rearrange the columns in descending order based on their column values, moving from left to right ("4. Rearrange columns).

In the last step, start from the beginning and recalculate the row values using the binary transformation. In the example below, the numerical values of the rows are sorted in the correct order ($14 > 12 > 11 > 3$). The iteration is now finished. However, if this recalculation results in a different order of the rows, the entire algorithm must be repeated (second iteration).

This procedure is also known as the "rank order clustering algorithm" and has been developed by J. R. King back in 1980.

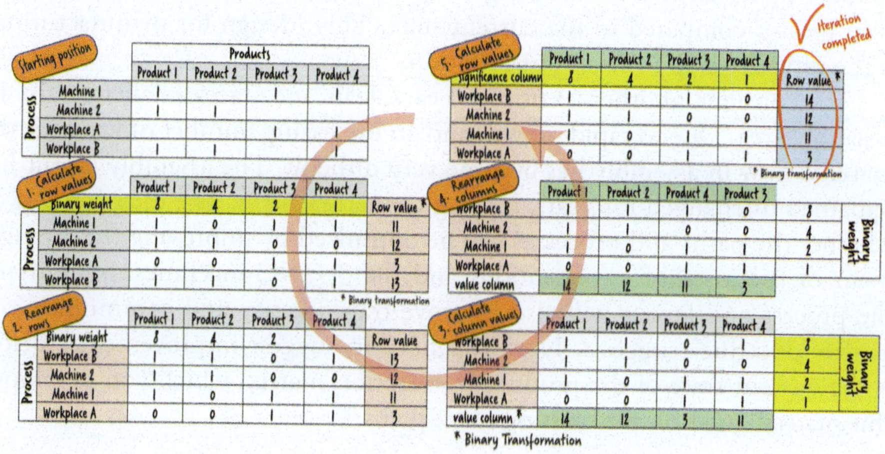


What does this mean for the organization of your new production cells? The goal is to group machines according to the zoning, thereby significantly minimizing distances and improving the flow of information within the department. It turns out that two possible cells (cell "orange" and cell "pink") are the most optimized arrangement of machines and workstations. However, despite this optimization, not all products can be produced in the indicated cells. For example, product 1 must also be placed on machine 1 in cell "pink" and product 3 must also be placed in cell "orange" for an operation. Nevertheless, the unbundling of the material flows is significant compared to the initial situation.

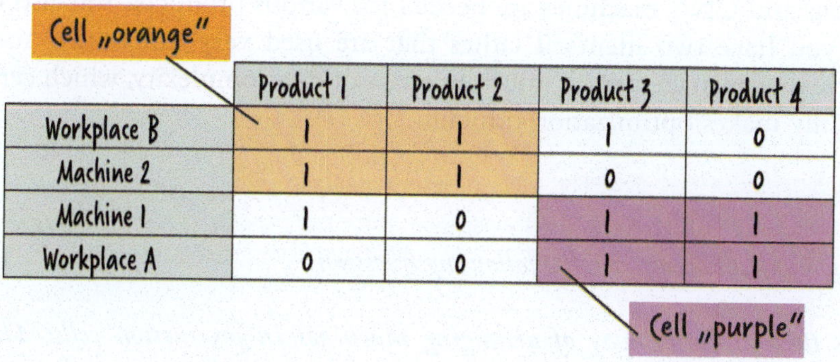

This new arrangement of machines can also lead to changes in product design or processes over time, such that, for example, the operation of product 1 on machine 2 is no longer necessary. This makes the simplification even greater compared to the current state. This "design for manufacturing" is the ultimate step in Lean optimization.

Let's consider another example: LeanClean Inc. faces challenges in its assembly area. Due to rapid growth and an increasing number of variants, the material flow in assembly has become very difficult. The assembly layout has remained unchanged for years, leading to inefficiencies. The question now is whether the paths and processes can be optimized by improving the arrangement of the assembly workstations and the 3 CNC machines. Simplifying the process and layout will also improve transparency and communication, as the "doctrine" suggests. Now you have to suggest improvements. Below is the current state of the product workstation matrix, which you developed this matrix based on the existing work plans.

| | Products | | | | | | | |
|---|---|---|---|---|---|---|---|---|
| | Octopus "Alles-Könner" | Octopus "Bodenexperte" | Octopus "Wassersauger" | Elephant "Economy" | Elephant "Maxi" | Snake EU | Snake USA | Snake RoW |
| Assembly workplace 1 | 1 | 1 | 1 | | | | | |
| Assembly workplace 2 | 1 | 1 | | 1 | | | 1 | |
| Assembly workplace 3 | | | | | 1 | 1 | 1 | |
| Assembly workplace 4 | 1 | 1 | 1 | | | | | |
| Assembly workplace 5 | 1 | 1 | 1 | 1 | | | | 1 |
| Assembly workplace 6 | | | | | | 1 | 1 | 1 |
| Assembly workplace 7 | | 1 | | | | | | |
| Gluing robot | | | | | 1 | 1 | 1 | 1 |
| Lathe 300 | 1 | | | 1 | 1 | 1 | | 1 |
| Lathe 300 | 1 | | | | | | | |
| Milling center | | 1 | 1 | 1 | 1 | 1 | 1 | |

At first glance, the product-workstation matrix does not allow you to recognize any improvement opportunities. It becomes clear that many workstations and CNC machines are needed for various products. You also notice that you have two identical lathes that are used very differently. You also believe that management is not fully aware of this complexity, which certainly certainly makes optimization difficult.

### *Task*

*You would like to answer the following questions:*

1. *Is there a better way of arranging machines in production cells? The goal is to unbundle material flows and simplifyy simplify responsibilities. Create the optimal arrangement/sequence of machines using the rank order clustering algorithm.*

2. *Suggest possible production cells and justify your decisions. Are there any other changes to the process or machinery as a result?*

## *Solution*

1. *The following optimal sequence of assembly workstations results from the rank order clustering algorithm after performing the first 3 steps of the first iteration:*

| | Products | | | | | | | | | |
|---|---|---|---|---|---|---|---|---|---|---|
| | Octopus "Alles-Könner" | Octopus "Bodenexperte" | Octopus "Wassersauger" | Elephant "Economy" | Elephant "Maxi" | Snake EV | Snake USA | Snake RoW | Sorted by row value | Binary weight |
| Assembly workplace 5 | 1 | 1 | 1 | 1 | | | | 1 | 241 | 1024 |
| Assembly workplace 1 | 1 | 1 | 1 | | | | | | 224 | 512 |
| Assembly workplace 4 | 1 | 1 | 1 | | | | | | 224 | 256 |
| Assembly workplace 2 | 1 | 1 | | 1 | | | 1 | | 210 | 128 |
| Lathe 300 | 1 | | | 1 | 1 | 1 | | 1 | 157 | 64 |
| Lathe 300 | 1 | | | | | | | | 128 | 32 |
| Milling center | | 1 | 1 | 1 | 1 | 1 | 1 | | 126 | 16 |
| Assembly workplace 7 | | 1 | | | | | | | 64 | 8 |
| Gluing robot | | | | | 1 | 1 | 1 | 1 | 15 | 4 |
| Assembly workplace 3 | | | | | 1 | 1 | 1 | | 14 | 2 |
| Assembly workplace 6 | | | | | | 1 | 1 | 1 | 7 | 1 |
| Binary weight | 128 | 64 | 32 | 16 | 8 | 4 | 2 | 1 | | |
| Value column calculated | 2016 | 1944 | 1808 | 1232 | 86 | 87 | 151 | 1093 | | |

To be rearranged by decreasing column value

*The intermediate result above shows the situation after step 3; the products must be sorted by their column values in step 4.*

| | | Products | | | | | | | | | |
|---|---|---|---|---|---|---|---|---|---|---|---|
| | | Octopus "Alles-Könner" | Octopus "Bodenexperte" | Octopus "Wassersauger" | Elephant "Economy" | Snake RoW | Snake USA | Snake EV | Elephant "Maxi" | Sorted by row value | Binary weight |
| Process | Assembly workplace 5 | 1 | 1 | 1 | 1 | 1 | | | | 241 | 1024 |
| | Assembly workplace 1 | 1 | 1 | 1 | | | | | | 224 | 512 |
| | Assembly workplace 4 | 1 | 1 | 1 | | | | | | 224 | 256 |
| | Assembly workplace 2 | 1 | 1 | | 1 | | 1 | | | 210 | 128 |
| | Lathe 300 | 1 | | | 1 | 1 | | 1 | 1 | 157 | 64 |
| | Lathe 300 | 1 | | | | | | | | 128 | 32 |
| | Milling center | | 1 | 1 | 1 | | 1 | 1 | 1 | 126 | 16 |
| | Assembly workplace 7 | | 1 | | | | | | | 64 | 8 |
| | Gluing robot | | | | | 1 | 1 | 1 | 1 | 15 | 4 |
| | Assembly workplace 3 | | | | | | 1 | 1 | 1 | 14 | 2 |
| | Assembly workplace 6 | | | | | 1 | 1 | 1 | | 7 | 1 |
| | Binary weight | 128 | 64 | 32 | 16 | 1 | 2 | 4 | 8 | | |
| | Sorted by value column | 2.016 | 1.944 | 1.808 | 1.232 | 1.093 | 151 | 87 | 86 | | |

*In the final step of the first iteration, you will notice that the order of the rows has changed, requiring you to run the entire algorithm a second time. After the second iteration, there is no further change in the order of the columns. At that point you stop the process, having found the optimal allocation of processes to products.*

| Process | Octopus "Alles-Könner" | Octopus "Bodenexperte" | Octopus "Wassersauger" | Elephant "Economy" | Snake RoW | Snake USA | Snake EV | Elephant "Maxi" | Sorted by line value (2nd Iteration) | Binary weight |
|---|---|---|---|---|---|---|---|---|---|---|
| Assembly workplace 5 | 1 | 1 | 1 | 1 | 1 | | | | 248 | 1024 |
| Assembly workplace 1 | 1 | 1 | 1 | | | | | | 224 | 512 |
| Assembly workplace 4 | 1 | 1 | 1 | | | | | | 224 | 256 |
| Assembly workplace 2 | 1 | 1 | | 1 | | 1 | | | 212 | 128 |
| Lathe 300 | 1 | | | 1 | 1 | | 1 | 1 | 155 | 64 |
| Lathe 300 | 1 | | | | | | | | 128 | 32 |
| Milling center | | 1 | 1 | 1 | | 1 | 1 | 1 | 119 | 16 |
| Assembly workplace 7 | | 1 | | | | | | | 64 | 8 |
| Gluing robot | | | | | 1 | 1 | 1 | 1 | 15 | 4 |
| Assembly workplace 6 | | | | | 1 | 1 | 1 | | 14 | 2 |
| Assembly workplace 3 | | | | | | 1 | 1 | 1 | 7 | 1 |
| Binary weight | 128 | 64 | 32 | 16 | 8 | 4 | 2 | 1 | | |
| Column value (2. Iteration) | 2.016 | 1.944 | 1.808 | 1.232 | 1.094 | 151 | 87 | 85 | | |

→ Sequence of products according to the value column (2nd iteration) is correct (descending order) – therefore the algorithm must be terminated (optimum found)

2. *You have developed two alternative solutions: The first solution includes the two cells "green" and "blue" (see circle 1 and circle 2). It is still debatable whether two lathes are needed. You have decided to sell one of the two identical lathes to save space. You have also made a conscious decision to assign the milling center and assembly workstation 7 to the green cell (see circle 3). This allows you to produce the Octopus "Bodenexperte", the Octopus "Wassersauger", and the Elephant "Economy" all in the green cell. If you had assigned these two processes to the blue cell, this would not have been possible. The three products would have to go through both cells.*

| Process | Octopus "Alles-Könner" | Octopus "Bodenexperte" | Octopus "Wassersauger" | Elephant "Economy" | Snake RoW | Snake USA | Snake EV | Elephant "Maxi" |
|---|---|---|---|---|---|---|---|---|
| Assembly workplace 5 | 1 | 1 | 1 | 1 | 1 | | | |
| Assembly workplace 1 | 1 | 1 | 1 | | | | | |
| Assembly workplace 4 | 1 | 1 | 1 | | | | | |
| Assembly workplace 2 | 1 | 1 | | 1 | | 1 | | |
| Lathe 300 | 1 | | | 1 | 1 | | 1 | 1 |
| ~~Lathe 300~~ | 1 | | | | | | | |
| Milling center | | 1 | 1 | 1 | | 1 | 1 | 1 |
| Assembly workplace 7 | | 1 | | | | | | |
| Gluing robot | | | | | 1 | 1 | 1 | 1 |
| Assembly workplace 6 | | | | | 1 | 1 | 1 | |
| Assembly workplace 3 | | | | | | 1 | 1 | 1 |

*The second solution also has two cells "green" and "blue". You now keep both lathes, and each cell gets a lathe of identical design. In order for the two products Octopus "Alleskönner" and Elephant "Economy" to be produced completely in the green cell, the work plan must be changed, and the process adapted so that the other lathe is being used (see circle A). You also reduce the complexity for all products that pass through the "blue" cell, as the "blue" cell now also has an integrated lathe. The Snake EU and Elephant "Maxi" can now be produced entirely in the "blue" cell.*

| Process | Octopus "Alles-Könner" | Octopus "Bodenexperte" | Octopus "Wassersauger" | Elephant "Economy" | Snake RoW | Snake USA | Snake EU | Elephant "Maxi" |
|---|---|---|---|---|---|---|---|---|
| Assembly workplace 5 | I | I | I | I | I | | | |
| Assembly workplace 1 | I | I | I | | | | | |
| Assembly workplace 4 | I | I | I | | | | | |
| Assembly workplace 2 | I | I | | I | | I | | |
| Lathe 300 | X | | | X | I | | I | I |
| Lathe 300 | I (A) | | | I | | (B) | | |
| Milling center | | I | I | I | | I | I | I |
| Assembly workplace 7 | | I | | | | | | |
| Gluing robot | | | | | I | I | I | I |
| Assembly workplace 6 | | | | | I | I | I | I |
| Assembly workplace 3 | | | | | | I | I | I |

*As you can see, the products Snake RoW and Snake USA still have to go through both cells. Therefore, you have decided to discuss a modified production process with the engineering department: the two products in question would have to be modified so that they could be assembled at assembly workstations 6 or 3 in the "blue" cell. You thought this would be an easy change, but the different assembly cells are equipped with very specific tools and test equipment, so that in addition to the process change, minor new investments are also required. So, it stays exciting...*

## 3.3 Method 11: Single Minute Exchange of Die (SMED)—Reducing the Setup Time is Important but Challenging

A few years ago, LeanClean Inc. purchased three CNC automatic lathes for complex metal parts. Since the technology is still new to the company, the setup process can be greatly improved. Therefore, a SMED project was initiated to reduce setup times. SMED is a method designed to optimize changeover process through both organizational and technical measures. Previously, LeanClean successfully applied this method to reduce setup times on the Plastmaster 2000.

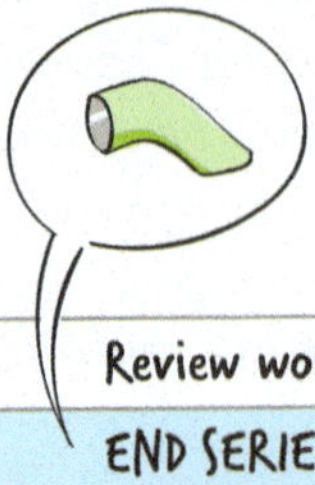

| Phase | No. | Step | Time (min) | Int. mandatory? | External possible? |
|---|---|---|---|---|---|
| Preparation (external) | 1 | Review work plan | 8 | | |
| | 2 | END SERIES "NOZZLE GREEN" | - | | |
| Setting up During Shutdown (internal) | 3 | Get granules from the warehouse | 10 | | |
| | 4 | Get tool in store | 10 | | |
| | 5 | Clean tool | 7 | | |
| | 6 | Provide open-end wrenches 8,10,15 | 3 | | |
| | 7 | Allen wrench missing, borrow | 5 | | |
| | 8 | Wait until crane is free | 10 | | |
| | 9 | Remove tool with crane | 12 | | |
| | 10 | Lift new tool with crane | 5 | | |
| | 11 | Position and tighten | 15 | | |
| | 12 | Screw on hoses (one defective) | 10 | | |
| | 13 | Get replacement hose from store | 8 | | |
| | 14 | Mount spare hose | 6 | | |
| | 15 | Start up machine | 10 | | |
| | 16 | Produce and check test parts | 5 | | |
| | 17 | Adjust, produce test parts | 5 | | |
| | 18 | Release production in ERP | 3 | | |
| Postprocessing (external) | 19 | START SERIES "HANDLE SHELL ORANGE" | - | | |
| | 20 | Fill out Q-document | 7 | | |

| Total time required in min | |
|---|---|
| Internal | 124 |
| External | 15 |

You will recall the central basic approach of SMED: the distinction between "internal setup" and "external setup" (see section 4.4 in "LEAN

Production - Simple and Comprehensive"). "Internal setup", refers to activities where the machine is idle, while "external setup" includes tasks performed while the machine is still processing a job. After the initial setup recording, the activities are categorized. The goal is to identify and convert as many "internal" processes into "external" ones, minimizing machine downtime. In this way, the internal setup time of the Plastmaster 2000 setup process was reduced from 124 to 68 min. Then try to reduce individual waste in the remaining internal setup processes, using automation to reduce the final time.

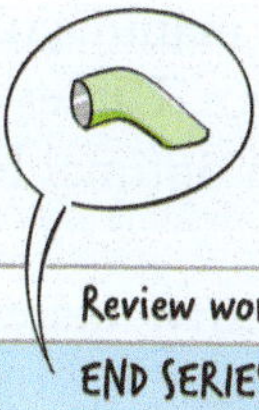

| Phase | No. | Activity | Time (min) | Int. mandatory? | External possible? |
|---|---|---|---|---|---|
| Preparation (external) | 1 | Review work plan | 8 | | |
| | 2 | END SERIES "NOZZLE GREEN" | - | | |
| Setting up During Shutdown (internal) | 3 | Get granules from the warehouse | 10 | | X |
| | 4 | Get tool in store | 10 | | X |
| | 5 | Clean tool | 7 | | X |
| | 6 | Provide open-end wrenches 8,10,15 | 3 | | X |
| | 7 | Allen wrench missing, borrow | 5 | | X |
| | 8 | Wait until crane is free | 10 | | X |
| | 9 | Remove tool with crane | 12 | | |
| | 10 | Lift new tool with crane | 5 | | |
| | 11 | Position and tighten | 15 | | |
| | 12 | Screw on hoses (one defective) | 10 | | |
| | 13 | Get replacement hose from store | 8 | | X |
| | 14 | Mount spare hose | 6 | | |
| | 15 | Start up machine | 10 | | |
| | 16 | Produce and check test parts | 5 | | |
| | 17 | Adjust, produce test parts | 5 | | |
| | 18 | Release production in ERP | 3 | | X |
| Postprocessing (external) | 19 | START SERIES "HANDLE SHELL ORANGE" | - | | |
| | 20 | Fill out Q-document | 7 | | |

| Total time required in min | | New: After check (X) "external possible" - Total time required in min. |
|---|---|---|
| Internal | 124 | → 68 |
| External | 15 | → 71 |

Now let's closely examine the exercise example. You have performed a setup recording for the CNC lathes mentioned earlier: armed with a stopwatch and a writing pad, you documented all relevant activities of representative setup operations for several days. After this detailed recording, you brought the entire CNC production team together to review and discuss the findings, focusing on understanding the observed operations and brainstorming improvement measures.

The table below shows you the results of the setup time recording of a part on the lathe 345 of the "CNC part small round brass". The brass part measures 80 mm in length and 20 mm in diameter. It is a relatively small part that has some complexity and requires a high level of manufacturing accuracy. The manufacturing cost is currently €735.20 per piece. Under current conditions, this part is set up 10 times a year and with an optimal lot size is 1510 pieces.

The expensive cost is due to the part's complexity, material properties (special alloy), and various outsourced operations such as coating and heat treatment.

| | | Start time | End time | Time required | Date | Comment |
|---|---|---|---|---|---|---|
| | 1 Prepare Setup – Get setup papers | 08.40 | 08.55 | 00.15 | 10/11 | |
| | 2 Dismantle reducing tube | 14.50 | 14.57 | 00.07 | 13/11 | |
| | 3 Maintenance machine – reducing tube clogged | 14.57 | 17.00 | 02.03 | 13/11 | No maintenance for 2 months |
| | 4 Install reducing tube, set up chuck | 17.00 | 17.10 | 00.10 | 13/11 | |
| | 5 Prepare tools on mounting trolley | 17.10 | 17.40 | 00.30 | 13/11 | |
| | 6 Tool presetting and measuring | 17.40 | 18.10 | 00.30 | 13/11 | |
| | 7 Align pipe holder | 18.10 | 18.13 | 00.03 | 13/11 | |
| | 8 Load program, import tool data | 18.13 | 18.45 | 00.32 | 13/11 | |
| | 9 Personal allowance time employee | 18.40 | 19.00 | 00.20 | 13/11 | |
| | 10 Install external turret 1/align tools | 19.00 | 19.20 | 00.20 | 13/11 | |
| Setup (S) | 11 Install cutting inserts in circular saw drill rev. 1 | 19.20 | 19.45 | 00.25 | 13/11 | Quick clamps missing |
| | 12 Install HS + GS collet chucks | 19.45 | 19.53 | 00.08 | 13/11 | |
| | 13 Prepare tools for internal turret 2 | 19.53 | 20.07 | 00.14 | 13/11 | |
| | 14 Install/align tools internal turret 2 | 20.07 | 20.22 | 00.15 | 13/11 | |
| | 15 Personal allowance time employee | 20.22 | 20.27 | 00.05 | 13/11 | |
| | 16 Tool calibration on external turret 1 | 20.27 | 21.20 | 00.53 | 13/11 | |
| | 17 Tool calibration on internal turret 2 | 21.20 | 21.44 | 00.24 | 13/11 | |
| | 18 Clean up workplace | 21.44 | 21.50 | 00.06 | 13/11 | |
| | 19 Replenish bar loader during order processing | 21.50 | 21.55 | 00.05 | 14/11 | Not clear what material |
| | 20 Employee has other tasks | 06.00 | 07.00 | 01.00 | 14/11 | other job has priority |
| | 21 Machine dry run – geometry & drainage check | 07.00 | 08.00 | 01.00 | 14/11 | Is it necessary? Simulation |
| | 22 Meeting – employee not at machine | 08.00 | 08.20 | 00.20 | 14/11 | |
| | 23 Maintenance machine – hollow spindle clogged | 08.20 | 09.10 | 00.50 | 14/11 | |
| | 24 Personal allowance time employee | 09.10 | 09.30 | 00.20 | 14/11 | |
| | 25 Running in 1st part in brass | 09.30 | 10.30 | 01.00 | 14/11 | Production 1st test part |
| Run-in (RI) | 26 Adjust coolant nozzles | 10.30 | 11.15 | 00.45 | 14/11 | |
| | 27 Personal allowance time employee | 11.15 | 11.30 | 00.15 | 14/11 | |
| | 28 Measuring parts clamped in the chuck | 11.30 | 12.30 | 01.00 | 14/11 | |
| | 29 Personal allowance time employee, lunch break | 12.30 | 13.15 | 00.45 | 14/11 | |
| | 30 Get pipe material for order in warehouse | 13.15 | 14.00 | 00.45 | 14/11 | |
| | 31 Shift handover, handover meeting | 14.00 | 14.30 | 00.30 | 14/11 | |
| | 32 Correct part dimensions | 14.30 | 17.30 | 03.00 | 14/11 | |
| | 33 Machine problems at the start of production | 18.00 | 22.00 | 04.00 | 14/11 | Time recording corrrected (30' missing) |
| | 34 Production of 5 parts and release parts after check | 06.00 | 09.30 | 03.30 | 15/11 | 3 hours for Q release |
| | P Start production – production parts | 09.30 | ... | 44.00 | 4 days | Production went well |
| | A Dismantle for next job – clean up | 06.30 | 08.00 | 01.30 | 21/11 | |

Legend: 07.45 means 7 hours and 45 minutes

The entire team was very surprised when the recorded times were added up: an unbelievable 27 h and 55 minutes of machine downtime to complete the setup, run-in, and dismantling. LeanClean Inc. divides the changeover process into the time components "actual setup", "run-in", and "dismantle". However, all three phases are often referred to in abbreviated form as "setup".

| Total time required for individual steps | | |
|---|---|---|
| Setup (S) | 7.25 hours = 445 minutes | 27 hours und 55 minutes |
| Run-in (RI) | 19.00 hours = 1140 minutes | |
| Dismantle (D) | 1.30 hours = 90 minutes | |
| Production (P) | 44 hours = 2640 minutes | |

During the discussion, when these setup costs were compared with the actual production time of 44 hours, it became clear to the entire team that urgent action was required.

***Task***

1. *In the initial situation, all activities under "setup", "run-in", and "dismantle" are internal setup operations. This means the machine remains idle during these tasks. To improve this we need to analyze the list of activities from number 1 to 34 and determine which of each activity can be turned into "external setup" in the future. In addition, name the possible improvements per activity number.*
2. *What is the improvement in setup time, including run-in and dismantling, based on your suggestions—in absolute values and in %? What cost savings result based on an hourly rate of €120?*
3. *What is the new optimum lot size after the setup time has been reduced? Today, the lot size is 1,510 pieces. However, for this calculation, you must know some additional data: the hourly rate is €120, the annual demand of the part "CNC part small round brass" is 15,000 pieces per year, and the average storage cost h was determined as €44.11 /piece for one year. This is calculated from the storage interest rate p = 6% and a unit price c = €735.20 (h = p × c).*

**Tip** To answer the question of the optimal lot size $Q_{opt}$, it is best to use the well-known Andler formula. This formular shows that the optimal lot size significantly depends on the setup cost A, along with other influencing factors such as the annual demand D and the annual storage cost of the parts. The result of the formula gives you an initial orientation for an economically justified lot size. However, the formula simplifies in many respects and neglects various negative side effects of large lot sizes like the resulting inventory and longer lead times. Therefore, anyone who uses Andler for their calculations should use SMED.

$$Q_{opt} = \sqrt{\frac{2 \times A \times D}{p \times c}}$$

| | |
|---|---|
| $Q_{opt}$ | Lotsize |
| D | Yearly demand |
| A | Setup cost per setup process |
| $p \times c$ | Storage cost h of the part per year with p = storage interest rate and c = cost of the part (unit price) |

## *Solution*

1. *Activity numbers 1, 3, 5, 6, 8, 10, 11, 16, 18, 19, 20, 22, 24, 30, 31, and 33 can be converted into external setup with relatively simple measures. "External" here means that the CNC machine produces parts while the operator performs these activities.*

   *As you can see, many measures are purely organizational in nature and can be implemented without major investment. it is important to note that personal allowance times, such as break periods, should not be reduced, as they are also considered part of setup process.*

| | No. | Task | Time required in hrs min | Internally mandatory? | External possible? | Comment |
|---|---|---|---|---|---|---|
| Setup (S) | 1 | Prepare Setup – Get setup papers | 00.15 | | X | Prepare documents before job end |
| | 2 | Dismantle reducing tube | 00.07 | X | | |
| | 3 | Maintenance machine – reducing tube clogged | 02.03 | | X | Adhere to maintenance windows |
| | 4 | Install reducing tube, set up chuck | 00.10 | X | | |
| | 5 | Prepare tools on mounting trolley | 00.30 | | X | Preparing the tools |
| | 6 | Tool presetting and measuring | 00.30 | | X | Prepare tools |
| | 7 | Align pipe holder | 00.03 | X | | |
| | 8 | Load program, import tool data | 00.32 | | X | Prepare / test program offline |
| | 9 | Personal allowance time employee | 00.20 | X | | |
| | 10 | Install external turret 1/align tools | 00.20 | | X | Set up while machine is still running |
| | 11 | Install cutting inserts in circular saw drill rev. 1 | 00.25 | | X | Investment additional circular saw |
| | 12 | Install HS + GS collet chucks | 00.08 | X | | |
| | 13 | Prepare tools for internal turret 2 | 00.14 | X | | |
| | 14 | Install/align tools internal turret 2 | 00.15 | X | | |
| | 15 | Personal allowance time employee | 00.05 | X | | |
| | 16 | Tool calibration on external turret 1 | 00.53 | | X | Setup while machine is still running |
| | 17 | Tool calibration on internal turret 2 | 00.24 | X | | |
| | 18 | Clean up workplace | 00.06 | | X | Can be done while machine is still running |
| | 19 | Replenish bar loader during order processing | 00.05 | | X | Set up material during order processing |
| Run-in (RI) | 20 | Employee has other tasks | 01.00 | | X | Improve organization / leadership |
| | 21 | Machine dry run – geometry & drainage check | 01.00 | X | | |
| | 22 | Meeting – employee not at machine | 00.20 | | X | Improve organization / leadership |
| | 23 | Maintenance machine – hollow spindle clogged | 00.50 | X | | |
| | 24 | Personal allowance time employee | 00.20 | | X | Possibilities for breaks when machine is running |
| | 25 | Running in 1st part in brass | 01.00 | X | | |
| | 26 | Adjust coolant nozzles | 00.45 | X | | |
| | 27 | Personal allowance time employee | 00.15 | X | | |
| | 28 | Measuring parts clamped in the chuck | 01.00 | X | | |
| | 29 | Personal allowance time employee, lunch break | 00.45 | X | | |
| | 30 | Get pipe material for order in warehouse | 00.45 | | X | Set up material while machine is still running |
| | 31 | Shift handover, handover meeting | 00.30 | | X | Improve organization / leadership |
| | 32 | Correct part dimensions | 03.00 | X | | |
| | 33 | Machine problems at the start of production | 04.00 | | X | Adhere to maintenance window |
| | 34 | Production of 5 parts and release parts after check | 03.30 | X | | |
| | P | Start production – production parts | 44.00 | | | |
| | A | Dismantle for next job – clean up | 01.30 | X | | |

Legend: 07.45 means 7 hours and 45 minutes

2. *By applying the time savings of 12 h and 34 min to the hourly rate, you achieve a saving of €1508 per setup—a significant cost reduction. However, the time saved could also be used to increase production capacity. With approximately 10 setups per year, the annual savings amount to €15,080.*

| | Existing process | | Setup time reduction - internal to external | | Potential new process |
|---|---|---|---|---|---|
| Setup (S) | 7.25 hours | | 05.39 hours | | 01.46 hours |
| Run-in (RI) | 19.00 hours | | 06.55 hours | | 12.05 hours |
| Dismantle (D) | 1.30 hours | | - hours | | 1.30 hours |
| | 27.55 hours | − | 12.34 hours | = | 15.21 hours |

3. *The optimal lot size after setup time optimization can now be set to 1120 pieces. The calculated lot size before setup time reduction was significantly higher at 1510 pieces. For this expensive part, a further reduction in lot size would be desirable. However, these additional improvements now require technical optimizations, such as additional tools, fixtures, and automation. Moreover, in many cases, it is only by using of these setup aids that the reduction to the "single digit minutes" (SMED) can be achieved.*

$$Q_{opt} = \sqrt{\frac{2 \times A \times D}{p \times c}} = \sqrt{\frac{(2 \times \text{EUR}1842 \times 15,000\, pieces}{44.1 \frac{€}{pieces}}} \approx 1120\, pieces$$

$$A_{neu} = (15\, hours\ 21\, minutes) * €120\ per\ hour = €1842\, setup\ cost$$

## 3.4 Method 12: Lean Shelf for Optimized Arrangement of Parts Supply

The LeanClean company is setting up a new assembly cell for the production of a new model of the Elephant. Central to this is the lean shelf, designed not only to ensure an efficient assembly, but also to organize the flow of materials in such a way that prevents waste. As always in LeanClean's assembly, it is a given that the material flow logistics will be organized as a Kanban control loop. The lean shelf is a central element in the implementation of Kanban, and also an integral part of Kanban.

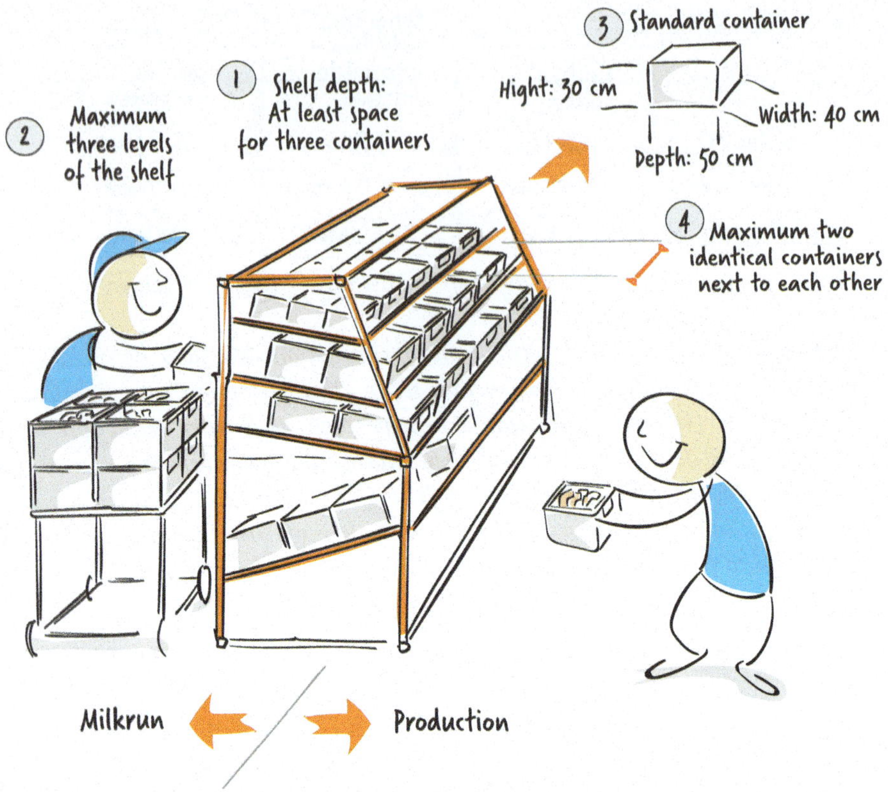


Based on your experience from designing previous shelves, you've set the following goals for the rack and the Kanban loop used to replenish the shelves.

- There are at least 2 containers per item on each shelf (2-bin Kanban).
- A shelf must be deep enough for 3 containers.
- There may only be a maximum of 2 containers with the same material next to each other (lane of 2 containers). This requirement results from the intensive experience with lean shelves at LeanClean: If too many containers with the same material are kept on the rack, maintaining FIFO becomes difficult, and it creates confusion for assembly.
- The standard container for this assembly has dimensions of 40 cm width, 30 cm height, and 50 cm depth. The goal is to use only one type of container for this assembly cell.

- The maximum height of the shelf should not exceed 1.80 m; therefore, it is planned with 3 shelves.
- High-runner parts should be stored next to the rack (stacking these items on the floor in a designated area has proven to be very useful). For the rack, parts requiring 7 or more bins on a shelf will not be placed on the rack.
- The rack is filled from the rear by the intralogistics system using a milkrun. The milkrun is expected to pass the assembly cell once a day at 9:00 AM.
- This rack is designed without a return shelf for the empty containers. Instead, the empty containers are collected in a special area next to the assembly cell.

To design the lean shelf for this assembly cell, you've received a table from the project manager. It includes data for each part: the daily consumption quantity $V_{max}$ and the number n of parts that fit into the desired standard container (determined through trial and error by an intern). Then, in the last column, there is the replenishment time of the parts in the various storage locations from which the milkrun draws each day. But is this information already sufficient for determining $t_{max}$?

| Part No. | Name | Consumption quantity per day (max.) | Max. number of parts per container | Replenishment time in days |
|---|---|---|---|---|
| 100001 | Kettle | 4 | 4 | 2 |
| 100002 | Contact socket | 4 | 6 | 2 |
| 100003 | Flange center | 4 | 67 | 2 |
| 100004 | Grid | 8 | 12 | 2 |
| 100005 | Cover round | 8 | 8 | 2 |
| 100006 | Metal cover | 24 | 21 | 2 |
| 100007 | Metal housing | 4 | 27 | 3 |
| 100008 | Flange right | 4 | 45 | 3 |
| 100009 | Cable set long | 5 | 13 | 3 |
| 100010 | Ladder - electric | 2 | 6 | 3 |
| 100011 | Holder input | 6 | 15 | 3 |
| 100012 | Electronics with plug | 89 | 100 | 3 |
| 100013 | Gasket round - large | 4 | 23 | 3 |
| 100014 | Support disc | 31 | 56 | 3 |
| 100015 | Plug with holder | 5 | 13 | 2 |
| 100016 | Base with hole | 5 | 13 | 2 |
| 100017 | Molded part specific | 26 | 22 | 2 |
| 100018 | Bolt long | 82 | 75 | 2 |
| 100019 | Protection plastics | 7 | 17 | 2 |
| 100020 | Bolt short | 5 | 45 | 4 |
| 100021 | Sheet metal thick | 4 | 10 | 4 |
| 100022 | Retaining plate short | 4 | 10 | 4 |
| 100023 | Spacer standard | 14 | 45 | 4 |
| 100024 | Spacer short | 11 | 67 | 4 |
| 100025 | Frame | 5 | 20 | 1 |
| 100026 | Gasket round - large | 5 | 35 | 1 |
| 100027 | Bracket right | 11 | 26 | 1 |
| 100028 | Spacer bush | 74 | 75 | 1 |
| 100029 | Cap large | 59 | 50 | 1 |
| 100030 | Cap round | 59 | 50 | 1 |

***Task***

1. *First, calculate the number of Kanban containers per article. Due to the "shaky" delivery capacity of some suppliers, you calculate with a safety factor of 2 at the first attempt.*

**Tip** The Kanban formula tells you how many Kanban cards or Kanban containers you need for a complete Kanban control cycle, starting from the supply process to the point of consumption (here: the lean shelf), so that you can always provide parts. The later exercise on Kanban in Sect. 3.6 expands on this formula.

However, it is shown here so that you can quickly determine how many containers you need per item for the Kanban cycle:

$$K = 1 + \frac{V_{max} \times t_{max}}{n} \times S$$

| | |
|---|---|
| K | Number of Kanban cards or containers |
| $V_{max}$ | Maximum consumption rate of a process or a product |
| $t_{max}$ | Maximum replenishment time or "restocking time" for a container or a "Kanban lot" |
| n | Number of parts in a container or per "Kanban lot" |
| S | Safety factor for the entire process ≥ 1. |

The Kanban formula gives you the number of cards (or containers). It is important to note that the Kanban parts can be either physically on the rack, in the milkrun (back or forward), or at the replenishing location. We always size the rack so that there is enough space for the maximum container quantity.

2. *Due to the specification of a maximum of 2 lanes with the same article, it must be examined whether the rack can accommodate all articles at all. Which of the items can you not accommodate on the shelves? Does this make sense? What happens if you reduce the safety factor to 1.4?*
3. *Consider how to position each item on the rack based on the safety factor of 1.4. Experience at LeanClean has shown that the middle shelf should be used for the "high runners", the bottom shelf for the "medium consumption items", and the top shelf for the "low runners". Define the "Top", "Middle", and "Bottom" shelves for each part number. Ensure that the three shelves have the same number of bin lanes. Approximately how wide will the rack be?*

**Solution**

1. *When calculating the number of containers required for the Kanban control cycle, the maximum replenishment time ($t_{max}$) is the replenishment time of the respective items at the replenishment location (replenishment time) plus 1 day. The extra day is added because a maximum of one day must be waited for a container that becomes empty after the milkrun at 9:00 AM. Irrespective of this, you always round up to the next full container at the end of each calculation. In combination with the safety factor of 2, you should be "on*

*the safe side" when designing the lean rack, even in the uncertain ramp-up situation of the new assembly cell.*

2. *Since a maximum of three containers per row was specified for the rack, the number of containers is now divided by 3 to determine the number of rows per item. Once the number of rows is > 2, the item must be stored outside the rack. For example, the six part numbers 100001, 100005, 100006, 1000012, 100017, and 100018 will have to be stored outside the rack.*

| | | Safety factor = 2 | | | Safety factor = 1.4 | | |
|---|---|---|---|---|---|---|---|
| Part No. | Name | Number of containers | Lanes per Part No. (calculated) | Location | Number of containers | Lanes per Part No. (calculated) | Location |
| 100001 | Kettle | 7 | 3 | outside | 6 | 2 | on the shelf |
| 100002 | Contact socket | 5 | 2 | on the shelf | 4 | 2 | on the shelf |
| 100003 | Flange center | 2 | 1 | on the shelf | 2 | 1 | on the shelf |
| 100004 | Grid | 5 | 2 | on the shelf | 4 | 2 | on the shelf |
| 100005 | Cover round | 7 | 3 | outside | 6 | 2 | on the shelf |
| 100006 | Metal cover | 8 | 3 | outside | 6 | 2 | on the shelf |
| 100007 | Metal housing | 3 | 1 | on the shelf | 2 | 1 | on the shelf |
| 100008 | Flange right | 2 | 1 | on the shelf | 2 | 1 | on the shelf |
| 100009 | Cable set long | 5 | 2 | on the shelf | 4 | 2 | on the shelf |
| 100010 | Ladder - electric | 4 | 2 | on the shelf | 3 | 1 | on the shelf |
| 100011 | Holder input | 5 | 2 | on the shelf | 4 | 2 | on the shelf |
| 100012 | Electronics with plug | 9 | 3 | outside | 6 | 2 | on the shelf |
| 100013 | Gasket round - large | 3 | 1 | on the shelf | 2 | 1 | on the shelf |
| 100014 | Support disc | 6 | 2 | on the shelf | 5 | 2 | on the shelf |
| 100015 | Plug with holder | 4 | 2 | on the shelf | 3 | 1 | on the shelf |
| 100016 | Base with hole | 4 | 2 | on the shelf | 3 | 1 | on the shelf |
| 100017 | Molded part specific | 9 | 3 | outside | 6 | 2 | on the shelf |
| 100018 | Bolt long | 8 | 3 | outside | 6 | 2 | on the shelf |
| 100019 | Protection plastics | 4 | 2 | on the shelf | 3 | 1 | on the shelf |
| 100020 | Bolt short | 3 | 1 | on the shelf | 2 | 1 | on the shelf |
| 100021 | Sheet metal thick | 5 | 2 | on the shelf | 4 | 2 | on the shelf |
| 100022 | Retaining plate short | 5 | 2 | on the shelf | 4 | 2 | on the shelf |
| 100023 | Spacer standard | 5 | 2 | on the shelf | 4 | 2 | on the shelf |
| 100024 | Spacer short | 3 | 1 | on the shelf | 3 | 1 | on the shelf |
| 100025 | Frame | 2 | 1 | on the shelf | 2 | 1 | on the shelf |
| 100026 | Gasket round - large | 2 | 1 | on the shelf | 2 | 1 | on the shelf |
| 100027 | Bracket right | 3 | 1 | on the shelf | 3 | 1 | on the shelf |
| 100028 | Spacer bush | 5 | 2 | on the shelf | 4 | 2 | on the shelf |
| 100029 | Cap large | 6 | 2 | on the shelf | 5 | 2 | on the shelf |
| 100030 | Cap round | 6 | 2 | on the shelf | 5 | 2 | on the shelf |

*By reducing the safety factor from 2 to 1.4, all items can now fit in the rack. However, you run a higher risk of a "production stoppage due to missing parts in the rack". Regarding the purchased parts, you accept this risk and inform the purchasing department about it. The purchasing department will now discuss with the suppliers to find solutions. In any case, you will personally supervise the implementation of the Kanban system.*

3. *With a safety factor of 1.4, you need 115 containers and 47 lanes of containers for the different part numbers. Since the rack requires three shelves of equal*

*size, 2 × 16 lanes and 1 × 15 lanes of containers can be accommodated per shelf. The order of consumption of the items ("descending by quantity") now determines the shelf for each item:*

| Part No. | Name | Number of containers | Lanes per Part No. (calculated) | Level |
|---|---|---|---|---|
| 100001 | Kettle | 6 | 2 | Middle |
| 100005 | Cover round | 6 | 2 | Middle |
| 100006 | Metal cover | 6 | 2 | Middle |
| 100012 | Electronics with plug | 6 | 2 | Middle |
| 100017 | Molded part specific | 6 | 2 | Middle |
| 100018 | Bolt long | 6 | 2 | Middle |
| 100014 | Support disc | 5 | 2 | Middle |
| 100029 | Cap large | 5 | 2 | Middle |
| 100030 | Cap round | 5 | 2 | Bottom |
| 100002 | Contact socket | 4 | 2 | Bottom |
| 100004 | Grid | 4 | 2 | Bottom |
| 100009 | Cable set long | 4 | 2 | Bottom |
| 100011 | Holder input | 4 | 2 | Bottom |
| 100021 | Sheet metal thick | 4 | 2 | Bottom |
| 100022 | Retaining plate short | 4 | 2 | Bottom |
| 100023 | Spacer standard | 4 | 2 | Bottom |
| 100028 | Spacer bush | 4 | 2 | top |
| 100010 | Ladder - electric | 3 | 1 | top |
| 100015 | Plug with holder | 3 | 1 | top |
| 100016 | Base with hole | 3 | 1 | top |
| 100019 | Protection plastics | 3 | 1 | top |
| 100024 | Spacer short | 3 | 1 | top |
| 100027 | Bracket right | 3 | 1 | top |
| 100003 | Flange center | 2 | 1 | top |
| 100007 | Metal housing | 2 | 1 | top |
| 100008 | Flange right | 2 | 1 | top |
| 100013 | Gasket round - large | 2 | 1 | top |
| 100020 | Bolt short | 2 | 1 | top |
| 100025 | Frame | 2 | 1 | top |
| 100026 | Gasket round - large | 2 | 1 | top |

16 lanes middle

16 lanes bottom

15 lanes top

115 containers

In total 47 lanes occupied, 1 lane still free

The rack will be at least 16 lanes × 40 cm = 640 cm wide. If you add the frame width of the rack profile of 2 × 4 cm and also a little spacing of 2 cm between the lanes and at the edge, the rack will be:

$640\text{cm} + (17 \times 2\text{cm}) + (2 \times 4\text{cm}) = 682\text{cm}\ \textit{wide}$

## 3.5 Method 13: The Milkrun—Takted Route Trains

In many companies, internal material handling is a series of individual trips with electric forklifts, like a cab carrying raw materials and other purchased parts to production; moving containers between production areas and machines, or moving finished parts to the warehouse. Occasionally, the shipping department simply says, "please find me the products so I can complete the order for the customer".

As a result, the warehouse and connecting aisles are often hectic and accident-prone. It doesn't help much if the forklift traffic is controlled by specific transport orders. The fact remains that the capacity of the forklift is limited to the small area of the forks, and you can expect 20 to 30% empty trips, and even up to 50% when unloading a truck.

A scenario at LeanClean Inc. will explicitly illustrate the problem. The figure below shows the detailed material flow relationships between the factory departments. With a few estimations and calculations, you can determine how cost-effective the milkrun concept is compared to forklift truck traffic.

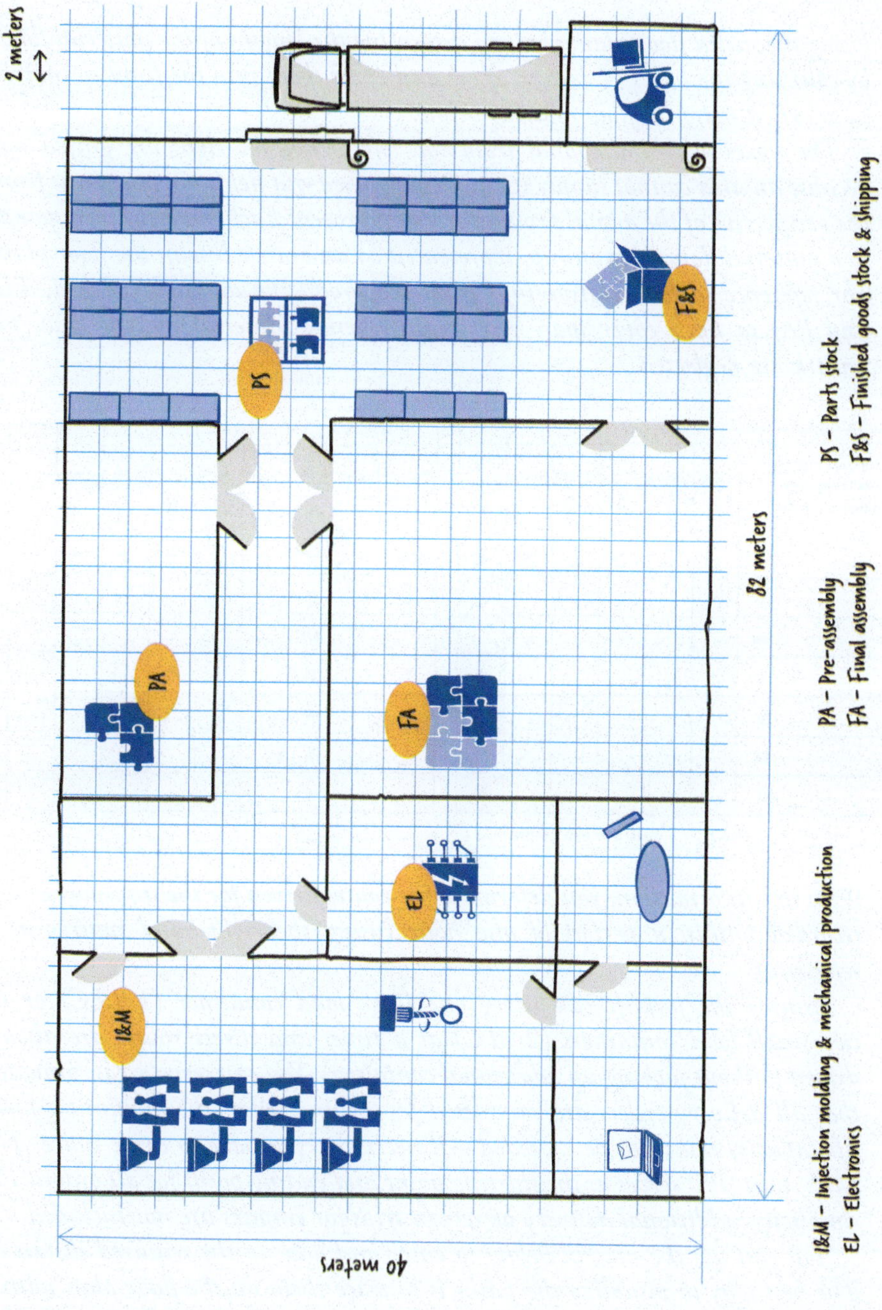
2 meters
82 meters
40 meters
PS
F&S
PA
FA
EL
I&M
I&M - Injection molding & mechanical production
EL - Electronics
PA - Pre-assembly
FA - Final assembly
PS - Parts stock
F&S - Finished goods stock & shipping

## *Task*

1. *Using the scaled floor plan, you can determine the lengths of the material flows by counting the boxes (1 box = 2 m) and enter them into the from-to matrix (also called material flow matrix) below.*

   *The boxes on the diagonal from top left to bottom right are grayed out because an area cannot supply itself. Other grayed-out fields are irrelevant from the perspective of the actual product creation process. Each free field corresponds to a material flow relationship, represented by an orange line in the floor plan. For reference, direct relationships from the producing areas (PS, I&M, EL, and PA) to F&S cover spare part requirements. Material returns flows for rework are neglected.*

| To / Standard — supply relationships with distances in meters / From | PS | I&M | EL | PA | FA | F&S |
|---|---|---|---|---|---|---|
| PS | | | | | | |
| I&M | | | | | | |
| EL | | | | | | |
| PA | | | | | | |
| FA | | | | | | |
| F&S | | | | | | |

For spare parts requirements

| To / Transport orders in time interval / From | PS | I&M | EL | PA | FA | F&S |
|---|---|---|---|---|---|---|
| PS | | 1 LLC 2 SLC | 1 SLC | 1 SLC | 2 SLC | |
| I&M | | | | 3 SLC | 1 LLC 2 SLC | 1 LLC |
| EL | | | | 2 SLC | 1 SLC | |
| PA | | | | | 4 SLC | 1 SLC |
| FA | | | | | | 1 LLC 2 SLC |
| F&S | | | | | | |

2. *With this information, you determine the routes driven for the transport orders recorded within a period of one hour (using the right-hand matrix as a reference).*

   *Assume the forklift can carry one large load container (1 LLC) or (!) two small load containers (2 SLC) at a time, and direct routes are always driven without picking up additional containers. You can serve areas, provided that all delivery orders are completed by the end. Also, the maximum forklift capacity must not be exceeded—if necessary, you have to drive twice. The parts stock (PS) serves as the starting point and ending point for all orders. For simplicity, additional delivery or return trips for empties are not included.*

   *You can use the matrix below to enter and add up the required distances. The best way to visualize the paths is to trace them on the floor plan with a pencil in your hand. You can then take the distances from the first matrix that you have filled out above. To illustrate the procedure, the distances for the first five transport orders are already entered in the matrix (of course, you could start with other orders).*

| To<br>Distances to be covered (with sequence)<br>From | PS | I&M | EL | PA | FA | F&S | Σ Row |
|---|---|---|---|---|---|---|---|
| PS | | 1) 3x46 m | 5) 46 m | | | | |
| I&M | | | | 2) 3x54 m | | | |
| EL | | | | | | | |
| PA | | | | | 3) 3 × 46 m | | |
| FA | | | | | | 4) 3 × 34 m | |
| F&S | 4)–5) 38 m | | | | | | |

Fields below the diagonal can be used for entering return paths

Total:

*So, how long is the total distance traveled in one hour, and how many individual trips with stops to pick up or drop off containers (transport cycles) do you count?*

3. *This admittedly quite elaborate exercise reflects well how much driving effort is associated with the many forklifts truck cyclces! Also note how much traffic there is in the central connecting corridor—you wouldn't want to be a pedestrian there. The planned milkrun should simplify many things.*

   *The route for the milkrun looks like the one shown on the next page. First, calculate the route length $s_R$ for one cycle from PS to PS by counting the boxes and compare the result with the total route length of the forklift traffic. What type of waste is there?*

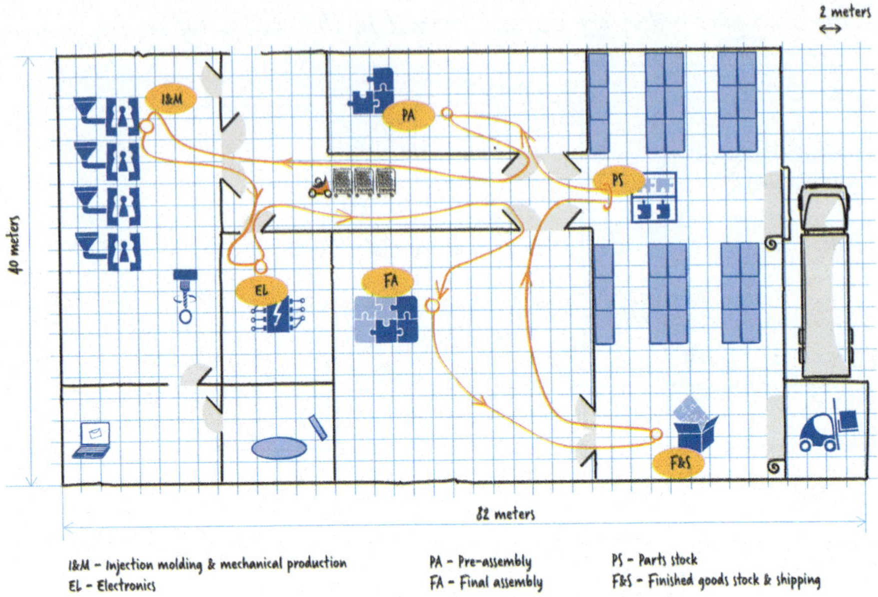


4. *But would the milkrun concept work in practice with the numerous requirements of internal material handling? To visualize the actual loading and unloading process at the stops, you simulate the loading situation of a route train with the transport orders for the forklift.*

   *To record the utilization situation of the trailers, we re-sort the from-to matrix shown at the beginning so that the rows and columns represent the transport orders in exactly the sequence of the milkrun route. This gives the new matrix at the bottom left. The orders above the diagonal represent the forward material flow to a later stop point on the route. The two orders below the diagonal are load containers that remain on the route train for the following cycle of the milkrun to be delivered in the next cycle (or the train would have to move backwards, which of course it does not do). To the right of the matrix, you can now enter, stop by stop, how many and which containers must be loaded and unloaded. Since full load containers are always exchanged for empty ones, the space on the trailer always remains occupied, even after an unloading process. For a better understanding of the procedure, the first two lines have already been filled out.*

| To / Transport orders in order of the stopping points of the route train / From (H) | 1. PS | 2. PA | 3. I&M | 4. EL | 5. FA | 6. F&S |
|---|---|---|---|---|---|---|
| 1. PS | | 1 SLC | 1 LLC<br>2 SLC | 1 SLC | 2 SLC | |
| 2. PA | | | | | 4 SLC | 1 SLC |
| 3. I&M | | 3 SLC | | | 1 LLC<br>2 SLC | 1 LLC |
| 4. EL | | 2 SLC | | | 1 SLC | |
| 5. FA | | | | | | 1 LLC<br>2 SLC |
| 6. F&S | | | | | | |

| Unloading in exchange for empties | Loading in exchange for empties | Places occupied on route train (full containers or empties) |
|---|---|---|
| Predecessor cycle | +1 LLC<br>+6 SLC | 1 LLC<br>6 SLC |
| 1 SLC | + 5 SLC | 1 LLC<br>11 SLC |
| | | |
| | | |
| | | |
| | Next cycle | |

*This little simulation exercise has given you a clear idea of what happens during a milkrun. The question that remains is how to design the capacity of the route train for the transport orders processed here. To answer this question, you can either add up all the entries for LLCs and SLCs in the from-to matrix on the left (this "value of the matrix" corresponds to the total turnover of load containers for the tour), or you can take the highest value of LLCs and SLCs on the route train in the right column of the simulation exercise.*

5. *To fully convince your colleagues of the feasibility of the milkrun, you determine the duration of a tour for the above number of transport orders. Simplifying VDI 5586 part 2, the average duration of a tour cycle $t_{zyk}$ can be determined from the time components for driving $t_F$, stopping $t_H$ and container change $t_W$ (loading, unloading) as follows:*

$$t_{zyk} = t_F + t_H + t_W$$

- *The travel time $t_F$ is obtained by dividing the route length $s_R$ already determined by the average route train speed $v_{RZ} = 6$ km per hour, which corresponds to 1.6 m per second.*
- *At the stops, there are time allowances, e.g., for acceleration and braking, for manual operations by the operator, and for getting on and off the tugger. At each of the stopping points $n_h = 6$ (i.e., including the warehouse) we assume a time loss of $t_h = 10$ s. By multiplication we obtain $t_H$.*
- *It is idealized to assume that each container change involves the exchange of a full container for an empty one in one direction or the other. However, for handling reasons, we distinguish between the time required for a LLC change ($t_{w,LLC} = 60$ s) and a SLC change ($t_{w,SLC} = 30$ s). From the simulation of the loading and unloading processes, it is easy to count in the*

*two columns in question how many containers of the LLC type and how many of SLC type have to be exchanged on the route under consideration. Adding the respective time components gives* $t_W$.

*How much time is the route train expected to take to complete the transport orders? What route train frequency do you propose for a test run phase?*

6. *Finally, you compare the cycle time of the route train that you have just determined, with the time required for the use of forklift trucks that has been practiced up to now. Here, you use the same formula as in the last calculation, whereby the forklift has a higher speed with* $v_{GSt} = 10$ *km/hour* = *2.74 m/second. The basis for the holding time calculation is* $t_h = 10$ *s. For the container change time* $t_{W,LLC}$ *we assume for simplification a change time of 60 s for each transport cycle, i.e. independent of the quantity, type, and loading condition of the exchanged containers. For the number of transport cycles, use the number of individual trips that you determined in task 2, as well as the total route length that the forklift must travel.*

**Tip** When determining the route frequency, first consider the transport volume and the required material replenishment times of the main processes to be connected by the route train. The material drop-off and pick-up times posted at the stops should be regular (e.g., every half hour or full hour) and reliable.

However, if the transport volume is greater than originally anticipated or increases for other reasons, the tugger train needs to be expanded by additional trailers. Other options include increasing the frequency of trips, adding a small and large loop, or even the use of an additional tugger train. All options have to be checked for their capacity feasibility and, of course, their economic viability.

Critical factors when sizing a route train are the resulting train length, the possible curve radiuses and turning areas, the track widths, and the allowable load and trailer load.

The dimensioning of the driving surfaces and the often-adjacent pathways must comply with the relevant workplace and health and safety regulations.

There is a wide range of trailer types that can be used for various load carriers and containers. Particularly heavy parts and large-volume load carriers and container racks must be placed on rollers for the convenience of logistics personnel. The trailers have special pick-up devices and ramp solutions for this purpose.

When implementing a milkrun concept, it is advicable to standardize the containers used in the plant.

You discuss responsibilities with the employees on site, and mark the pick-up and drop-off areas for materials and containers on the floor, or on selected shelves. You then post a timetable at each stop, just like at a bus station.

After a trial period—possibly even with borrowed equipment—an evaluation is made of what went well and what went badly, whether the capacity and timing of the tugger train met the requirements, and so on. Only then can an adapted concept be put into practice.

***Solution***

1. *Determined distances in the material flow matrix (your solution may be slightly different from the values shown here):*

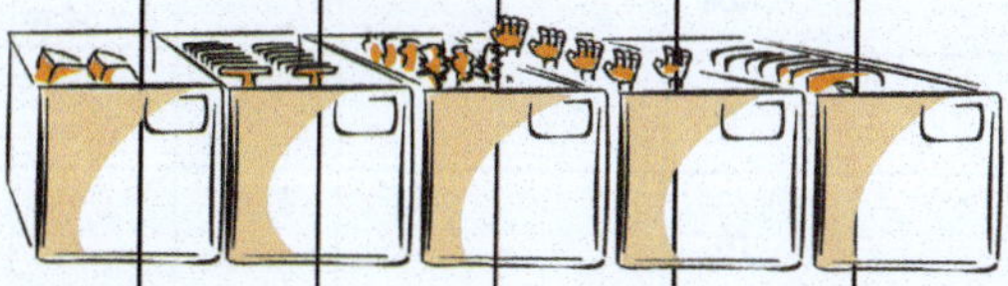

| To<br>Standard supply relationships with distances in meters<br>From | PS | I&M | EL | PA | FA | F&S |
|---|---|---|---|---|---|---|
| PS | | 46 | 46 | 24 | 26 | 38 |
| I&M | | | | 54 | 56 | 80 |
| EL | | | | 52 | 52 | 66 |
| PA | | | | | 46 | 54 |
| FA | | | | | | 34 |
| F&S | | | | | | |

For spare parts requirements

2. *The total distance is approximately 1400 meters with 29 trips (alternative sequences are possible).*

| To<br>Distances to be covered (with sequence)<br>From | PS | I&M | EL | PA | FA | F&S | Σ Row |
|---|---|---|---|---|---|---|---|
| PS | | 1)<br>3 × 46 m | 5)<br>46 m | 8)<br>24 m | 10)<br>26 m | | 234 meters |
| I&M | | | | 2)<br>3 × 54 m | 11)<br>3 × 56 m | 12)<br>80 m | 410 meters |
| EL | | | | 6)<br>52 m | 9)<br>52 m | | 104 meters |
| PA | | | 8)–9)<br>52 m | | 3)<br>3 × 46 m | 7)<br>54 m | 244 meters |
| FA | 9)–10)<br>38 m | 10)–11) 56 m<br>12)–11) 56 m | | | | 4)<br>3 × 34 m | 252 meters |
| F&S | 4)–5) 38 m<br>7)–8) 38 m<br>12)–PS) 38 m | | | | | | 114 meters |

m = meters

Fields below the diagonal represent the required connecting paths between the steps

Total: 1,358 meters

3. *A milkrun cycle from PS to PS: $s_R$ about 210 m. The forklift truck thus covers about 6.5 times the distance; waste type "transport". At this point, however, we are unsure whether a single tour can actually handle the transport orders that arise in one hour!*
4. *The tugger train needs to have a capacity of 4 LLCs and 21 SLLs for the transportation scenario. Of course, some "wiggle room" should be allowed for sizing in the context of a pilot test.*

| To<br>Transport orders in order of the stopping points of the route train<br>From | 1. PS | 2. PA | 3. I&M | 4. EL | 5. FA | 6. F&S | Unloading in exchange for empties | Loading in exchange for empties | <br>Places occupied on route train (full containers or empties) |
|---|---|---|---|---|---|---|---|---|---|
| 1. PS | | 1 SLC | 1 LLC<br>2 SLC | 1 SLC | 2 SLC | | Predecessor cycle | + 1 LLC<br>+ 6 SLC | 1 LLC<br>6 SLC |
| 2. PA | | | | | 4 SLC | 1 SLC | 1 SLC | + 5 SLC | 1 LLC<br>11 SLC |
| 3. I&M | | 3 SLC | | | 1 LLC<br>2 SLC | 1 LLC | 1 LLC<br>2 SLC | + 2 LLC<br>+ 5 SLC | 3 LLC<br>16 SLC |
| 4. EL | | 2 SLC | | | 1 SLC | | 1 SLC | + 3 SLC | 3 LLC<br>19 SLC |
| 5. FA | | | | | | 1 LLC<br>2 SLC | 1 LLC<br>9 SLC | + 1 LLC<br>+ 2 SLC | 4 LLC<br>21 SLC |
| 6. F&S | | | | | | | 2 LLC<br>3 SLC | Next cycle | Rest<br>5 full SLC for PA |

**Tip** Study order scenarios on a daily and weekly basis. The internal transport requirements can then be statistically recorded by type and frequency, creating a reliable database for sizing a tugger train.

5. *The cycle time of the milkrun is calculated as follows:*

$$t_{zyk} = \frac{210\ meters}{1.6\frac{meters}{seconds}} + 6 \times 10\ seconds$$
$$+ 8 \times 60\ seconds + 37 \times 30\ seconds$$
$$= 1781\ seconds \approx 30\ minutes$$

*The 5 full SLCs for pre-assembly (PA) remain on the tugger train for the next cycle; their replacement with empty containers is not included in this calculation.*

*For the pilot test, the milkrun could be run every 45 min or once an hour; this would give everyone involved enough time to familiarize themselves with the new processes; a 30-min start cycle would certainly be too ambitious.*

*Note: When calculating the tour cycle time, loading in the parts stock (PS) was treated as any other container change operation. This was done to facilitate the comparison with forklift traffic. Alternatively, a lump sum time could have been calculated for the uncoupling and coupling of already prepared trailers (see VDI 5586 sheet 2).*

6. *The time for the forklift is calculated as follows:*

$$t_{Fl.} = t_{F,Fl.} + t_{H,Fl.} + t_{W,Fl.}$$
$$= \frac{1400 meters}{2.4\frac{meters}{seconds}} + 29 \times 10\ seconds + 29 \times 60\ seconds$$
$$= 2613\ seconds$$
$$= 43.5\ minutes$$

*This means that the milkrun concept results in a time saving of 31% compared to forklift logistics. However, for a fair comparison of the workload, the additional time required to replace the remaining 5 SLCs would have to be included in the milkrun cycle time calculation, which would increase it by 2 min to 32 min. The calculated time savings would still be 26%. Add to this the numerous other benefits of the milkrun, and the use of a tugger train in a Lean environment should be pursued under all circumstances.*

*Note: Actual loading and unloading times are highly dependent on the design of the trailers and racks and, of course, the weight situation.*

## 3.6 Methode 14: Kanban—The Key Method for Achieving Pull

Kanban is certainly among the most relevant methods for creating flow in production. The basic approach is that a consumption process always "pulls" new parts from the production process when the corresponding buffer of parts at the point of consumption is slowly depleting. Kanban follows the pull principle. The pull principle is based on actual customer demand and controls production on that basis. In other words, if there is no customer demand, you are not allowed to produce! Thus, Kanban is a method, perhaps the most important of all, for generating "pull".

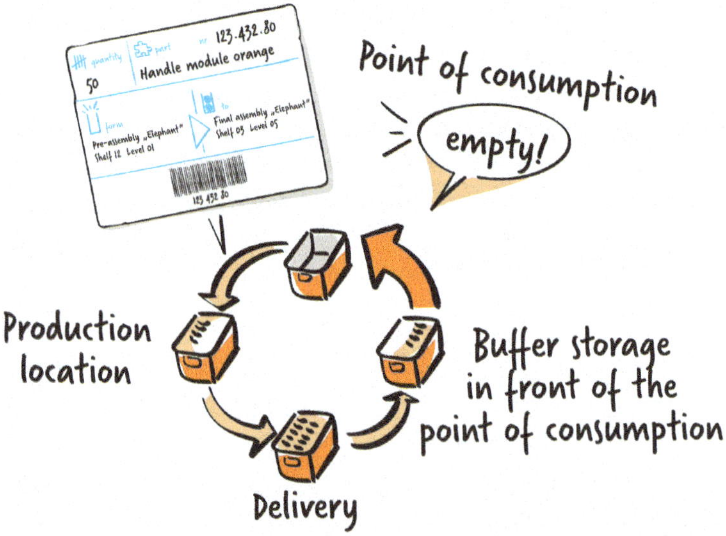


When establishing the Kanban control loop, we determine how many Kanban cards or containers needed in the cycle to ensure no missing parts at the point of consumption. Therefore, the two central components of the Kanban formula are the consumption rate of the parts at the destination on the one hand and the replenishment time on the other. In operations, the latter results from the time for the transmission of the Kanban signal (or the return of the Kanban container) to the production location, the waiting time until the production order is triggered (e.g., for batch size reasons), the production time, of course, and finally the transportation time of the parts to the consumption location.

The formula we use to calculate the Kanban quantity in this cycle is as follows:

$$K = 1 + \frac{V_{max} \times t_{max}}{n} \times S$$

| | |
|---|---|
| K | Number of Kanban cards or containers |
| $V_{max}$ | Maximum consumption rate of a process or a product in parts per minute |
| $t_{max}$ | Maximum replenishment time or "restocking time" for a container or a "Kanban Lot" in minutes |
| n | Number of parts in a container or per "Kanban Lot" (card or container) |
| S | Safety factor for the entire process $\geq 1$ (e.g., 1.5 at 50% safety factor) |

The following two exercises will help you get used to the formula and understand the effects of the variables involved.

**Kanban for Plastic Parts in Pre-assembly**

LeanClean Inc. currently works in pre-assembly in one shift, with injection-molded handle shells being required every 30 s. Under normal conditions, this assembly takt also represents the highest consumption rate $V_{max}$. The shift duration at LeanClean is 7:40 h.

The containers that move between pre-assembly and the injection molding machines are special containers with inserts that protect the parts from scratches. Each container has a capacity of n = 60 parts.

Since a milkrun is currently being introduced in production to supply parts to the operating areas, the milkrun cycle is an initial time component that must be considered when calculating the replenishment time. At the moment, the logistics specialist uses the route train to pick up the containers filled with injection molded parts twice per shift and take them to the various consumption locations. Due to the frequent retooling of the injection molding machines, the maximum waiting time for empty containers is only one shift until they are replenished. This is an excellent value—the Lean project for setup time reduction has done a great job.

Since LeanClean Inc. has limited experience with Kanban and the recently introduced milkrun concept, the safety factor is initially set to 50%. This is intended to "buffer" all imponderables.

**Tip** When designing a Kanban process, there are always ideas for further improvements. Examples include container standardization, inserts to protect parts from scratches or for better gripping during picking, and color-coding of storage, and pick locations ("5S"). The latter in particular can quickly induce minor or major layout optimizations. Consider all improvement ideas!

The result of dimensioning based on the Kanban formula is only a starting point for a functioning process. Once Kanban is functioning, it is easy to see where further improvements are necessary or possible. Optimizing Kanban is an iterative process whose goal is always to further reduce the number of parts in circulation (work in progress).

The formula is based on the idea that at least one container or Kanban is at the point of consumption and the others are in the replenishment loop. Of course, it is desirable that the parts consumption is as smooth as possible,—then Kanban works best. Instead of using the maximum values for the consumption V and the replenishment time t, average values could be used. However, using maximum values is a first step to protect the Kanban cycle against known fluctuations.

However, to be prepared for further irregularities in the process (e.g., in a ramp-up situation) and unforeseen fluctuations of unknown origin, a significant higher safety factor S is recommended. For example, an S = 2 doubles the number of Kanbans in the total cycle, which can be useful for initial sizing and for the first few weeks after the start of a new Kanban cycle. Especially in a pilot project (see Sect. 3.16), it is very important that "the thing works"—the minimum number of Kanbans can still be approached in the subsequent time.

***Task***

*Solve the following tasks:*

1. *Determine the minimum number of special containers required for the initial setup of the Kanban system.*
2. *Consider what uncertainties the safety factor compensates for. Think about the whole process including the information flow.*
3. *The resulting number of containers seems to be quite high to you. Which measures will lead to a sustainable reduction of the container quantity in the Kanban control loop?*
4. *You increase the number of parts in the container from the original 60 to n = 120 pieces. How does this change the total number of Kanban containers? What other aspects need to be considered when making such a change?*

***Solution***

1.

$$K_{\min} = 1 + \frac{2\frac{parts}{minute} \times (230 + 460 + 230) minutes}{60\frac{parts}{container}} \times 1.5$$
$$= 47\, containers$$

2. *All types of variability, such as delivery and quality problems, deviations and temporary failures in the milkrun process, temporarily increased parts consumption in pre-assembly, employee absences due to illness, and forecasting errors.*

3. *More frequent pick-ups and deliveries, even smaller batch sizes at the injection molding machines, higher part quality, more stable and reliable milkrun (justifying a smaller safety factor) and—formally speaking—larger containers (= pseudo solution, since no reduction in inventory is achieved), reduced variance in parts consumption.*
4. *The number of Kanban containers is reduced to 24. It is now necessary to discuss whether the advantages (fewer containers, less handling) outweigh the negative effects of the larger containers (container cost, newly dimensioned buffer space after the injection molding process and in pre-assembly, poorer ergonomics). Finally, this proposal must be seen in the desired standardization of containers.*

## Kanban for Metal Parts in a Milling Shop

The production manager of LeanClean Inc. has an explicit understanding of how to define the Kanban process for the handle shells. He also knows a nearby company, named the CNC-Exakt Ltd., which introduced Kanban years ago. He exchanges ideas with the company and discusses the following Kanban cycle in their milling shop:

The average demand for aluminum parts with the number 5822 of a group of six CNC milling machines is 1200 pieces per day in 2-shift operation. The working time of CNC-Exakt is 7:40 h per shift, as with LeanClean. The parts undergo several processing steps in the cutting and turning shop beforehand (collectively referred to as pre-fabrication) and are provided to the milling department in a central supermarket area for pickup by the milling shop workers. Since the machine locations of the areas are very close to each other, there is no need for intermediate material handling steps such as the milkrun at LeanClean. This is a significant difference to the process at LeanClean Inc.

Pre-fabrication is controlled via a Kanban lot-making board. On this board, all Kanban cards are collected in columns by part number. Only when a quantity deemed sufficient for manufacturing a certain part number has been reached ("economic lot size"), its production is triggered. As a result, it takes an average of three shifts until part 5822 is produced again. The production time of the then accumulated parts quantity is 600 min.

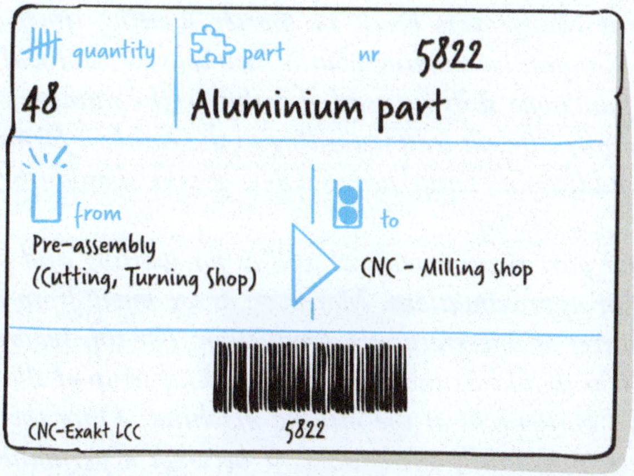


In contrast to the last exercise, this one use Kanban cards and generic standard containers. Each card represents a quantity of 48 parts and contains all logistics-relevant information, such as the place of origin and consumption. The 48 parts on a Kanban card are distributed among several small bins. Due to the proximity between the supplier and customer processes, alongside stable processes, CNC-Exakt Ltd. believes it does not need a safety factor.

### *Task*

*To check whether you have internalized the Kanban setup calculations correctly, solve the following tasks:*

1. *Determine the number of Kanban cards required to build the Kanban control loop of part number 5822. Instead of working with a maximum consumption, take the average values (but be aware of the risks involved...).*
2. *By which measure can the Kanban quantity be reduced further?*
3. *For what reasons should a safety factor still be included?*
4. *How would you physically arrange the Kanban containers between prefabrication and milling?*

### *Solution*

1. *The number of Kanban cards is (rounded up to the nearest whole number):*

$$K_{\min} = 1 + \frac{\frac{1200\,parts}{920\,minutes} \times (1380 + 600)\,minutes}{48\frac{parts}{Kanban\,cards}} \times 1$$
$$= 55\,Kanban\,cards$$

2. *More frequent changeovers result in shorter waiting times for the kanban cards on the Kanban lot-making board. Example: With a waiting time for lot formation of only one shift (instead of 3 shifts), the number of Kanban cards is reduced to only 30—at least according to the Kanban formula and without considering additional setup times. CNC-Exakt should therefore definitely have a SMED workshop.*
3. *The values for part requirements as well as for waiting and production times are averages or approximations. Moreover, as we know from experience, there are many sources of uncertainty in production. For this reason, a safety factor of at least 20% should be included in the initial setup of the Kanban control cycle to avoid bottlenecks at the milling machines. Once the Kanban control cycle has been successfully introduced, it becomes a permanent task ("CIP") to further stabilize the entire supply chain and gradually reduce the Kanban quantity.*
4. *A good solution would be to use a Lean shelf with inclined roller conveyors for the containers or a dedicated FIFO lane. Here, we discussed "lean shelf/rack" (see Sect. 3.4). This is a possibility, but certainly not a must. It is important that the space for the containers is planned, organized, and marked. It is not acceptable for the containers to be placed anywhere in production where space is available.*

## 3.7 Complementay Push Methods

Various exercises in this book will introduce you to many relevant approaches of the "Pull" methodology. You learn about Kanban, milkrun, sequencing, the lean shelf, or even balancing. All these approaches move in the direction of Pull and help you realize it. And that's exactly what we want - we want as much "Pull" in production as possible, because this significantly reduces waste or prevents it from occurring in the first place.

However, there are some parts or products that are not a perfect fit for the Pull approach. Either they are not sold in sufficient quantities, or they require a high level of custom engineering.

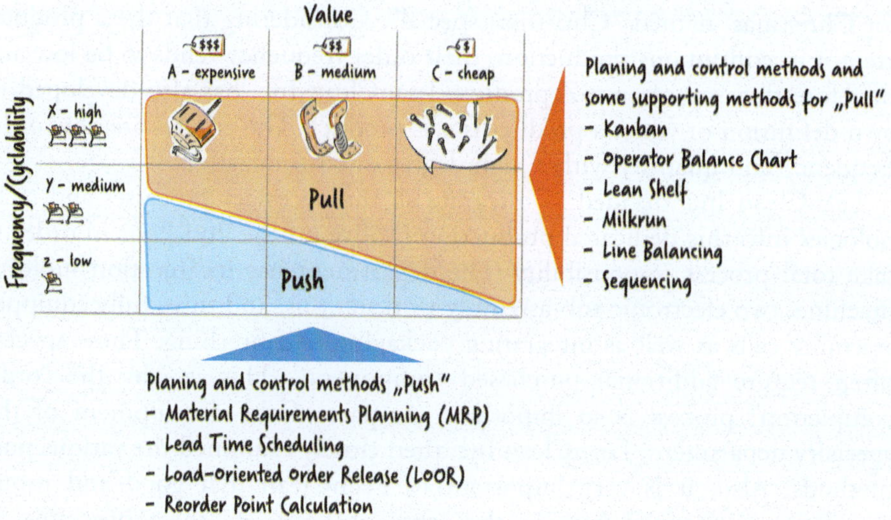


It is also important to understand that even in a Lean factory, "Push" planning and control approaches remain relevant. For example, LeanClean Inc. has a challenge with the well-known "Pull" methods for all specialty products, as certain materials "sit" in Kanban boxes for weeks, if not months, making it a waste. A Lean company can use push methods and still be a Lean company. However, push methods should be the exception, not the norm, and should only be used for special processes or products. The principle of "Pull before Push" always applies.

LeanClean has been discussing this topic very controversially lately and after much deliberation has decided to set up its own "special production".

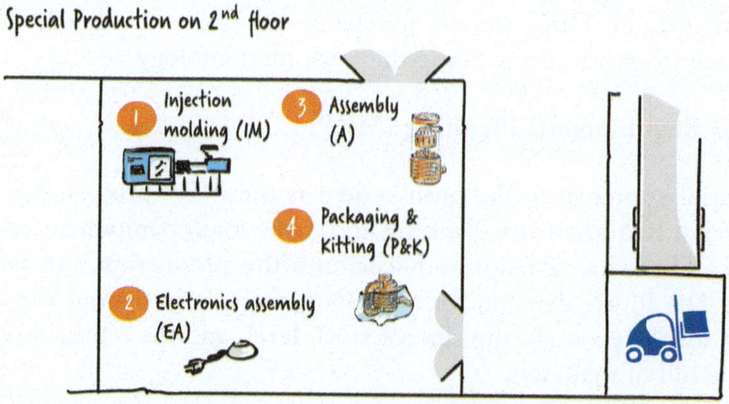


In this "special production", the vacuum cleaner models are produced in small quantities as well as vacuum cleaners upon special request by customers. For example, a customer may plan a campaign with a special vacuum cleaner for Christmas, named "Christmas special". Considering that these products are not in continuous production, their order frequency tends to be low and therefore they are often a Z product. LeanClean Inc. has also developed its own definition of when a product is a Z product. The criteria are: low order frequency and quantity with a high degree of customization.

LeanClean Inc. decided to integrate all the necessary processes and technologies into this dedicated production facility on the 2nd floor in order to take total process responsibility. The department has its injection molding machine, two electronic sub-assembly workstations, and three fully equipped assembly cells as well as integrating packaging and finishing. These specials often require additional purchased components. This is why the "order completion" process is so important. As part of the development of this specialty department, LeanClean Inc. then closely examined the various push methods. Also, it is very important to LeanClean that good and robust processes are mapped here in the sense of Lean. As these processes are supported by ERP, data quality must always be high. This is only possible if the processes are explicitly understood. The following four exercises will help you understand and apply the relevant push processes and procedures:

- Material Requirements Planning (MRP)
- Lead Time Scheduling
- Load Oriented Order Release (LOOR)
- Reorder Point Calculation.

These four push methods give you a good idea of the differences compared to pull. Of course, the complexity and capabilities of today's ERP solutions are much greater. However, the following exercises are still relevant and a good place to start understanding the push methodology.

### Material Requirements Planning (MRP)

The central approach in the push system is the MRP process. MRP stands for Material Requirements Planning and is the main component of any PPS software. This process plans and schedules the production and purchasing process. The future demand for a component is derived from the customer demand for the product, the current stock level, and the replenishment time using the bill of materials.

The MRP process goes as follows:

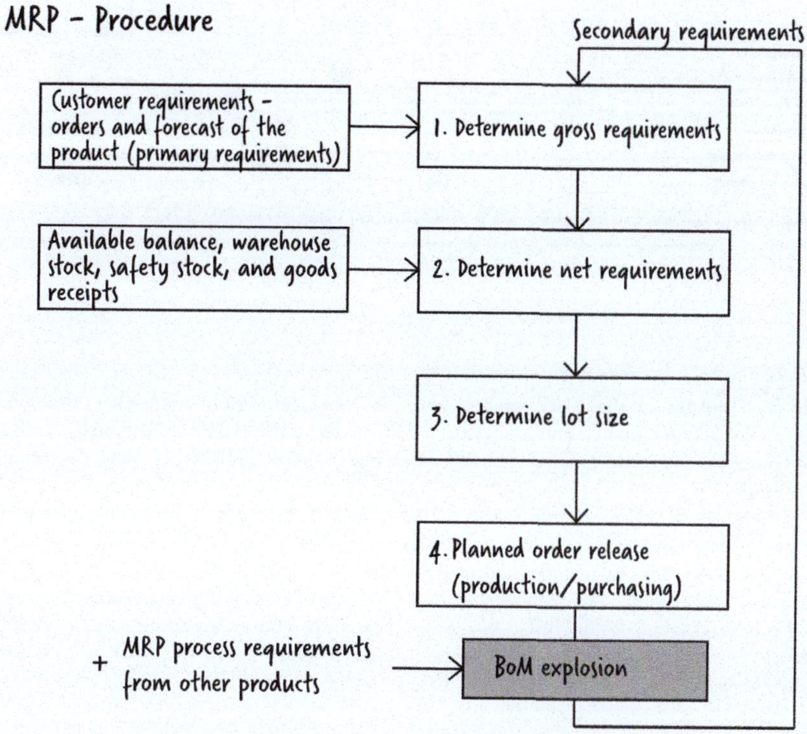


The gross demand is derived from the actual orders and sales forecast. This is then compared to the available inventory, which is calculated as current inventory minus the safety stock, but also includes all receipts. If this available inventory is less than the gross demand for the period, a production order is then planned for the components manufactured at LeanClean Inc. or a purchase order proposal is generated for purchased materials. The lot size is determined according to standard lot-sizing procedures (e.g., based on the Andler formula). After the BOM is exploded, the same process is performed for all assemblies and components. The resulting component-level requirements are known as dependent requirements. The following diagram illustrates the MRP process using concrete data.

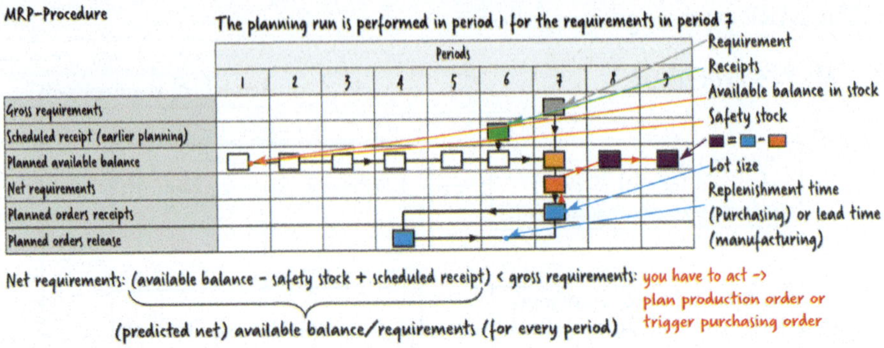


**Tip** The quality of the MRP process and the subsequent push approaches, depends on the quality of the master and transaction data in the ERP system. Therefore, administrative processes and data quality crucially determine if MRP process can fully contribute to planning.

LeanClean Inc. also recognized this. When they switched to custom manufacturing, a lot of master data had to be compiled newly. This process alone has made LeanClean Inc. a better company. Many bills of material had to be revised and updated. Keeping bills of material up to date is challenging, especially for custom products. These products are often customized and produced in small quantities. As a result, the effort required to create accurate BOMs is not always easy to justify. However, without accurate master data, the MRP process simply does not work.

Now let's examine the parts list for the "handle red standard".

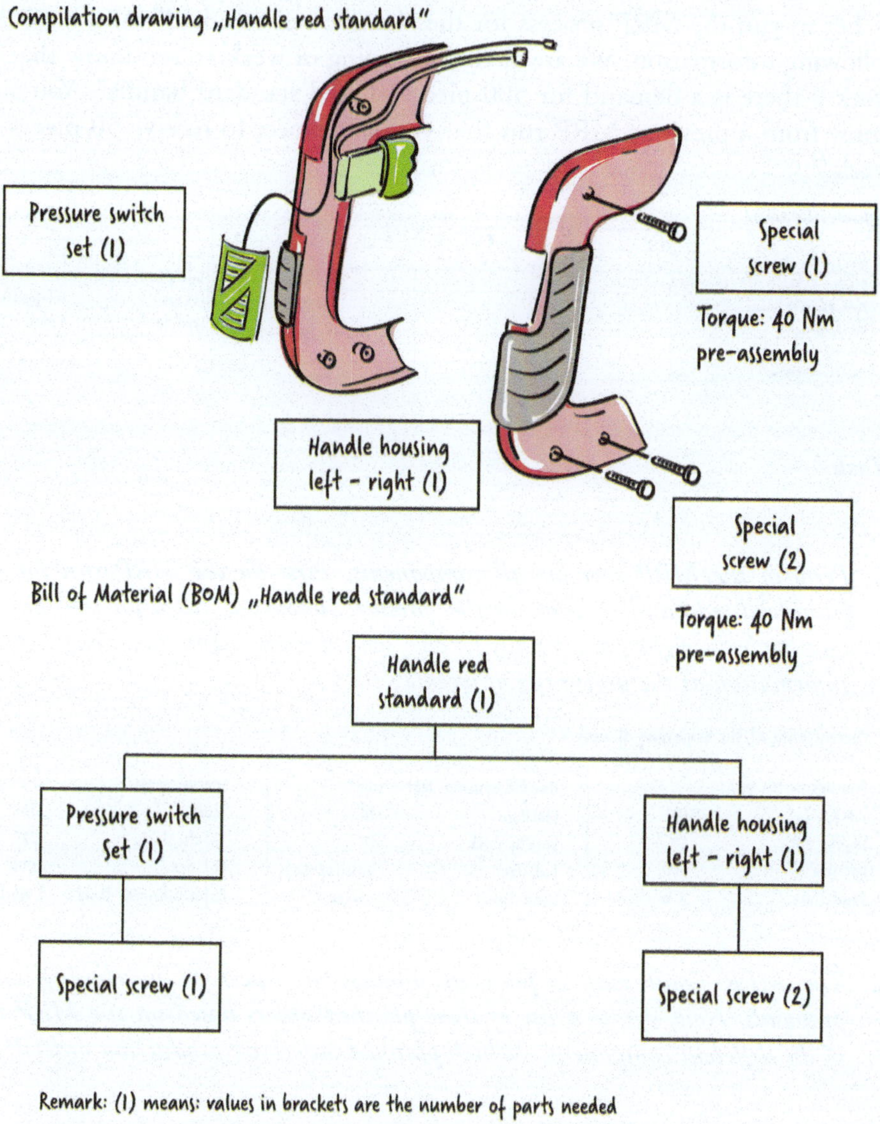


This bill of material will be complemented with different master data such as lot size method, safety stock, lead time, and replenishment time for every part.

Let us run the MRP process for the "handle red standard". You have the following information: We are at the beginning of week 1. You know that in week 9 there is a demand for 500 pieces of "red standard handle". You also know from a previous MRP run that you can expect to receive 50 pieces in week 7.

| Handle red standard | Week | | | | | | | | |
|---|---|---|---|---|---|---|---|---|---|
| | 1 | 2 | 3 | 4 | 5 | 6 | 7 | 8 | 9 |
| Gross requirements | | | | | | | | | 500 |
| Scheduled receipt | | | | | | | 50 | | |
| Planned available balance | | | | | | | | | |
| Net requirements | | | | | | | | | |
| Planned orders receipt | | | | | | | | | |
| Planned orders release | | | | | | | | | |

| | |
|---|---|
| Inventory | 360 |
| Safety stock | 40 |
| Lot size | Lot-for-lot |
| Lead time | 3 weeks |

## Task

1. *Perform the MRP run for all components. First do the MRP run for the "standard red handle", then for the "pressure switch set", then for the "handle housing left – right", and finally for the "special screws". You have the following information at the beginning of week 1.*

Current data at the beginning of week 1

| Pressure switch set: | | Handle housing left - right: | | Special srews: | |
|---|---|---|---|---|---|
| Stock | 120 | Stock | 20 | Stock | 180 |
| Safety stock | 50 | Safety stock | 20 | Safety stock | 120 |
| Lot size | 600 | Lot size | Lot-for-Lot | Lot size | 5000 |
| Lead time | 4 Wochen | Lead time | 1 week | Replenishment time | 1 Woche |

2. *The MPR result run is planned production orders or purchasing order proposals. Now, create a list of these planned orders based on the MRP run of the different components. Which planned orders are production orders?*

## Solution

1. *The MRP run for each component is as follows:*

Handle red standard

| | Week 1 | 2 | 3 | 4 | 5 | 6 | 7 | 8 | 9 |
|---|---|---|---|---|---|---|---|---|---|
| Gross requirements | | | | | | | | | 500 |
| Scheduled receipt | | | | | | | 50 | | |
| Planned available balance | 320 | 320 | 320 | 320 | 320 | 320 | 320 | 370 | 370 |
| Net requirements | | | | | | | | | 130 |
| Planned orders receipt | | | | | | | | | 130 |
| Planned orders release | | | | | | 130 | | | |

| | |
|---|---|
| Inventory | 360 |
| Safety stock | 40 |
| Lot size | Lot-for-lot |
| Lead time | 3 weeks |

*You now know that week 6 will be the start of production for 130 pieces of the* "handle red standard". *You also need the "handle housing left–right".*

Handle housing left - right

| | Week 1 | 2 | 3 | 4 | 5 | 6 | 7 | 8 | 9 |
|---|---|---|---|---|---|---|---|---|---|
| Gross requirements | | | | | | 130 | | | |
| Scheduled receipt | | | | | | | | | |
| Planned available balance | 0 | 0 | 0 | 0 | 0 | 0 | 0 | 0 | 0 |
| Net requirements | | | | | | 130 | | | |
| Planned orders available | | | | | | 130 | | | |
| Planned orders release | | | | | 130 | | | | |

| | |
|---|---|
| Inventory | 20 |
| Safety stock | 20 |
| Lot size | Lot for lot |
| Lead time | 1 week |

*... you also need enough "pressure switch sets"...*

Pressure switch set

| | Week 1 | 2 | 3 | 4 | 5 | 6 | 7 | 8 | 9 |
|---|---|---|---|---|---|---|---|---|---|
| Gross requirements | | | | | | 130 | | | |
| Scheduled receipt | | | | | | | | | |
| Planned available balance | 70 | 70 | 70 | 70 | 70 | 70 | 540 | 540 | 540 |
| Net requirements | | | | | | 60 | | | |
| Planned orders receipt | | | | | | 600 | | | |
| Planned orders release | | 600 | | | | | | | |

| | |
|---|---|
| Inventory | 120 |
| Safety stock | 50 |
| Lot size | 600 |
| Lead time | 4 weeks |

*The two modules "handle housing left - right" and "pressure switch set" require special screws for installation. These are stocked with the individual plastic parts for each set, ensuring that a sufficient number of special screws are available. Although this process is time-consuming, it is important to Lean-Clean Inc. Years ago, no products could be manufactured for a month because the special screws were missing.*

*The requirement for special screws comes from the MRP run of the two modules "handle housing left - right" and "pressure switch set" as well as the parts list information. Please note that the "handle housing left - right" requires 2 special screws each.*

Special screws

| | Week 1 | 2 | 3 | 4 | 5 | 6 | 7 | 8 | 9 |
|---|---|---|---|---|---|---|---|---|---|
| Gross requirements | | 600 | | | 260 | | | | |
| Scheduled receipt | | | | | | | | | |
| Planned available balance | 60 | 60 | 4460 | 4460 | 4460 | 4200 | 4200 | 4200 | 4200 |
| Net requirements | | 540 | | | | | | | |
| Planned orders available | | 5000 | | | | | | | |
| Planned orders release | 5000 | | | | | | | | |

| | |
|---|---|
| Inventory | 180 |
| Safety stock | 120 |
| Lot size | 5000 |
| Replenishment time | 1 week |

2. *The following planned orders result from the MRP runs of the various products, modules, and parts:*

| Order Number | Name | Start date | End date | Quantities |
|---|---|---|---|---|
| *10001* | *Handle red standard* | *Week 6* | *Week 9* | *130 pieces* |
| *10002* | *Handle housing left—right* | *Week 5* | *Week 6* | *130 pieces* |
| *10003* | *Pressure switch set* | *Week 2* | *Week 6* | *600 pieces* |
| *10004* | *Special screws* | *Week 1* | *Week 2* | *5000 pieces* |

*Order numbers 10001, 10002, and 10003 are production orders, while 10004 is a purchase order. Regarding production orders, the lead time is the same as the internal throughput time, while for purchasing order, the replenishment lead time is the time required to source the parts from the vendor plus the time required to do the incoming inspection.*

## Lead Time Scheduling

Lead time scheduling determines when a product is to be manufactured in-house and how the individual operations must be scheduled to meet the agreed-on customer deadlines. A distinction is made between forward scheduling (start date is determined by scheduling, assuming a theoretical lead time, empirical value) and backward scheduling (end date is derived from the customer deadline). The pure production time is called lead time. Lead time scheduling does not consider capacity constraints. Capacity scheduling is when capacity constraints are considered. The resulting dates result in capacity requirements, which you can be used for load-oriented order release (LOOR) calculations.

The basis for leadtime scheduling is the network we now use to schedule an order for a LeanClean Inc. product in the specialty department.

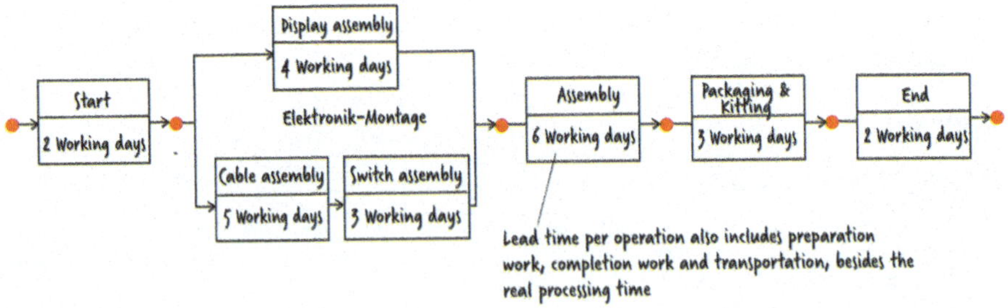


*Task*

1. *Schedule this job based on this network. Make a list of all the operations, including the earliest possible start and earliest possible finish of the task. The earliest possible start date for the entire order is set to November 10. What is the total lead time and when is the earliest possible finish date for the entire order? Therefore, the list must be completed as follows:*

| Work process | Earliest start date | Earliest end date |
|---|---|---|
| Start | 0 (10.11.) | 2 |
| Display assembly | 2 | 6 |
| ... | ... | ... |

*Notes: You can also enter the earliest start and end dates from "0–2" directly into the activity boxes in the network. When transferring the dates to the "real world", you need to include a calendar—after all, weekends are off.*

2. *Is there a critical path? Mark the critical path.*

**Tip** The critical path is the sequence of work steps in which no delay should occur. If there is a delay along the critical path, the end date is always pushed back by the length of the delay.

The exercise above is the forward scheduling method. You could also use "backward scheduling" logic. This involves starting with the customer's requested date and working backwards through the network to find the latest possible start and finish dates for each operation. This is how we find the latest feasible start date for the entire order.

The difference between the earliest and latest possible start date provides the scheduling department with a "time buffer" to release the work order.

***Solution***

1. *The following schedule can be displayed:*

| Working process | Earliest start date | Earliest end date |
|---|---|---|
| Start | 0 | 2 |
| Display assembly | 2 | 6 |
| Cable assembly | 2 | 7 |
| Switch assembly | 7 | 10 |
| Assembly | 10 | 16 |
| Packaging & Kitting | 16 | 19 |
| End | 19 | 21 |

Lead time in total 21 working days

*The lead time for the order is 21 working days, and if you start the order on 10.11., the order will be completed on 8.12. at the earliest (based on a 5-day week).*

| November | | | | | | |
|---|---|---|---|---|---|---|
| Mo | Di | Mi | Do | Fr | Sa | So |
| | | 1 | 2 | 3 | 4 | 5 |
| 6 | 7 | 8 | 9 | 10 | 11 | 12 |
| 13 | 14 | 15 | 16 | 17 | 18 | 19 |
| 20 | 21 | 22 | 23 | 24 | 25 | 26 |
| 27 | 28 | 29 | 30 | | | |

| Dezember | | | | | | |
|---|---|---|---|---|---|---|
| Mo | Di | Mi | Do | Fr | Sa | So |
| | | | | 1 | 2 | 3 |
| 4 | 5 | 6 | 7 | 8 | 9 | 10 |
| 11 | 12 | 13 | 14 | 15 | 16 | 17 |
| 18 | 19 | 20 | 21 | 22 | 23 | 24 |
| 25 | 26 | 27 | 28 | 29 | 30 | 31 |

Start date: 10.11.
End date: 8.12.

2. *The critical path for this order includes the path "Start - Cable Assembly - Switch Assembly - Assembly - Package and Finish – Finish". The display assembly can be delayed by 4 days and the customer's finish date can still be met.*

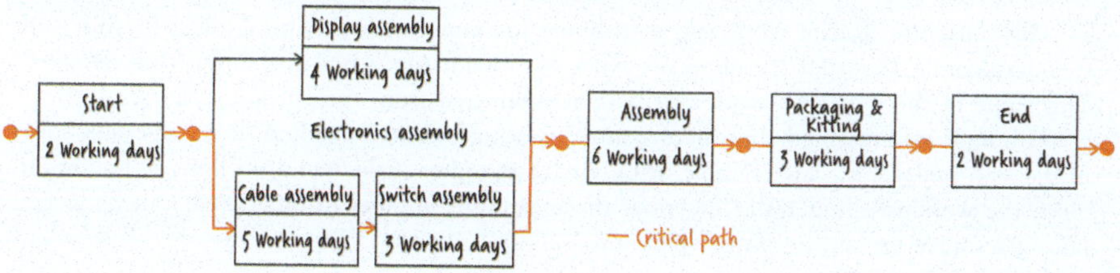


## Load-Oriented Order Release (LOOR)

A characteristic of this new "Specials" department is the quantity and high variability of sometimes small production batches. Too much planned capacity utilization, i.e., too many released production orders, can induce long queues in front of the workstations. The flow is disrupted, and lead times increase.

**Tip** One method of optimizing sequencing is load-oriented order release (LOOR). In this method, pending production orders are released according to the current capacity utilization of a shopfloor, and are then allocated to the workstations or production systems within the shopfloor based on the load values too. The criterion for the utilization of a production order is the load limit, which is defined by the utilization percentage.

The utilization percentage (EPS) is a theoretical value that is calculated as follows, based on the current capacity of the shop and considering the Load Barrier (BS):

$$\begin{aligned} Load\ Barrier\ BS &= average\ stock + planned\ out\ flow \\ &= target - stock + capacity \\ &= capacity \times EPS \end{aligned}$$

$$Thus: EPS\ in\ \% = \frac{BS \times 100\%}{capacity}$$

The utilization percentage (EPS) serves as a control parameter for the capacity utilization of a resource. In practice, the EPS is used at 200% to 300%. Utilization rate is a control parameter only and should not be confused with the available capacity.

In addition, the scheduling of later work steps is reduced, or their start date is moved to the future. It is generally assumed that later operations will probably not be completed in the period under consideration. Therefore, the full capacity requirements of these orders are devalued as they move further into the future.

$$For\ work\ step\ p = 1: Work\ Step\ Devaluation\ Factor\ ABFA(1) = 1$$

$$Generally\ applies: Work\ Step\ Devaluation\ Factor\ ABFA\ for\ work\ step\ p$$

$$ABFA(p) = \left| \left( \frac{100}{EPS} \right) \times (1 - p) \right|$$

The capacitive load of work step p in hours (or minutes) is then multiplied by this work step devaluation factor for the respective work step (load (p) x ABFA (p)) and is thus devalued. This value of the respective work step is then subtracted from the previous resulting load.

Although load-oriented shop floor control is a push control method, it incorporates some of the lean philosophy. For example, it only releases orders if the workload limit still allows it. Since the workload limit (WL) describes the maximum number of orders that can be loaded onto a resource group, the flow is planned.

LeanClean Inc. uses LOOR when releasing orders from the electronics assembly department (EA), the assembly department (A), and the packaging & kitting department (PK). This ensures that production is scheduled in the best possible sequence (in terms of capacity) with the available capacity (in this case, employee hours). The reality will probably be different, but the given sequence is a good starting point.

The utilization percentage (EPS) is constantly set to 200%. This results in the following step devaluation factors for the work processes:

| | Sequence of work step (work step p) | | | | |
|---|---|---|---|---|---|
| | p = 1 | p = 2 | p = 3 | p = 4 | p = 5 |
| Work Step Devaluation Factor ABFA | 1 | 0.5 | 0.25 | 0.125 | 0.0625 |

When releasing orders based on LOOR, the load limit plays an important role, as it describes the maximum possible utilization of orders for a work center and it is also the sum of the planned departures over a period plus the average level before the period. It is often colloquially referred to as the "available capacity of the work center"—which is not quite correct but easier to understand. In summary, the initial situation is as follows:

| | Electronics assembly (EA) | Assembly (A) | Packaging & Kitting (PK) |
|---|---|---|---|
| Capacity in hours | 60 | 60 | 40 |
| EPS in % | 200 | 200 | 200 |
| Load Barrier in hours | 120 | 120 | 80 |

Based on the throughput scheduling of the various orders, MRP has created a list of the sequence of planned production orders for the three departments ("work centers" in LOOR), including the planned load.

| Order number | Sequence of work step with workload in hours and in work areas | | | Target Start date | Sequence | Release remark |
|---|---|---|---|---|---|---|
| | 1 | 2 | 3 | | | |
| 101 | 30 EA | 50 A | 40 PK | 8. Dez. | 4 | |
| 102 | 20 PK | 30 EA | X | 5. Dez. | 2 | Spare part |
| 103 | 30 PK | 40 EA | 20 A | 6. Dez. | 3 | Spare part |
| 104 | 20 EA | 30 A | X | 2. Dez. | 1 | |
| 105 | 30 PK | 10 A | 20 EA | 12. Dez. | 6 | |
| 106 | 50 A | 10 PK | 30 EA | 13. Dez. | 7 | |
| 107 | 40 A | 40 PK | 20 EA | 10. Dez. | 5 | |

Legend: 30 EA means: 30 hours load for electronic assembly EA
1, 2, 3 means: Sequence of work steps, workstation sequence
X means: No step or sequence in the job

For order number 101, the sequence begins with electronics assembly (EA), followed by assembly (A), and is then completed in packaging & kitting (PK). In addition to the packaging and completion (P&C) work processes, the P&C work center, being the most specialized workstation in the department, has also specialized in special pre-assembly work over time. As a result, the entire department has become much more flexible, enabling synergies and fully utilizing employee skills.

As you can see, load-oriented order release is only used for the 3 work centers electronics assembly, assembly, and packaging & completion. The plastic injection process is scheduled using a separate production order.

### *Task*

*Perform a load-oriented job shop control for the 7 planned production orders. Note the remaining workloads in the following table and the uniform EPS of 200%, as used above to create the ABFA values. To do this, proceed line-by-line, i.e., take an order (according to the defined sequence) with all the associated work steps and load these efforts at the respective work centers.*

| Sequence | order number | Sequence of the work steps with the devaluated load in hours and in work areas | | | Release remark |
|---|---|---|---|---|---|
| | | EA (120) | A (120) | PK (80) | |
| | | Actual load | | | |
| | | 25 (95) | 0 (120) | 10 (70) | |
| 1 | 104 | | | | |
| 2 | 102 | | | | |
| 3 | 103 | | | | |
| 4 | 101 | | | | |
| 5 | 107 | | | | |
| 6 | 105 | | | | |
| 7 | 106 | | | | |

Legend:

Capacity: EA (120) = 120 hours of available capacity in the reporting period

Ausgangslage: 25(95) at work center EA means: 25 hours of capacity are required at EA to process the order in question. These 25 hours have been deducted from the total capacity (120) at this work center.
This means that only 95 hours of capacity remain at this work center.

25(95), 0(120), 10(70) -> The values in brackets are the (remaining) capacities of the respective stations, which should be used as a starting point for this task.

## *Solution*

*Orders are dispatched and debited according to their target start date from the throughput schedule.*

The orders are dispatched in the order of their urgency and devaluated according to the Work Step Devaluation Factor ABFA.

Load x Work Step Devaluation Factor = 20 x 1 = 20 (Residual capacities: 95 – 20 = 75)

30 x 0,5 = 15 (120 – 15 = 105)

20 x 0,25 = 5 (105 – 5 = 100)

| Sequence | order number | Sequence of the work steps with the devaluated load in hours and in work areas | | | Release remark |
|---|---|---|---|---|---|
| | | EA (120) | A (120) | PK (80) | |
| | | Actual load | | | |
| | | 25 (95) | 0 (120) | 10 (70) | |
| 1 | 104 | 20 (75) | 15 (105) | 0 (70) | |
| 2 | 102 | 15 (60) | 0 (105) | 20 (50) | |
| 3 | 103 | 20 (40) | 5 (100) | 30 (20) | |
| 4 | 101 | 30 (10) | 25 (75) | 10 (10) | |
| 5 | 107 | 5 (5) | 40 (35) | 20 (-10) ? | |
| 6 | 105 | - (5) | - (35) | 30 (-40) ? | |
| 7 | 106 | - (5) | 50 (-15) ? | 5 (-15) ? | |

? Load limit exceeded, as residual capacities < 0

*As you can see, there are three orders (105, 106, and 107) that have negative remaining capacity due to the planned load. In other words, these orders cannot be conducted in the specified period due to a lack of capacity.*

*A decision is now made to release each job based on the calculated load limits:*

| Sequence | Order number | Sequence of the work steps with the devaluated load in hours and in work areas | | | Release remark |
|---|---|---|---|---|---|
| | | EA (120) | A (120) | PK (80) | |
| | | Actual load | | | |
| | | 25 (95) | 0 (120) | 10 (70) | |
| 1 | 104 | 20 (75) | 15 (105) | 0 (70) | R |
| 2 | 102 | 15 (60) | 0 (105) | 20 (50) | R |
| 3 | 103 | 20 (40) | 5 (100) | 30 (20) | R |
| 4 | 101 | 30 (10) | 25 (75) | 10 (10) | R |
| 5 | 107 | 5 (5) | 40 (35) | 20 (-10) ① | R |
| 6 | 105 | - (5) | - (35) | 30 (-40) ② | nR |
| 7 | 106 | - (5) | 50 (-15) ① | 5 (-15) ② | nR |

Legend: R stands for „Release", nR stands for „no Release"

*If a work center's load limit has not been exceeded (or has only been exceeded with the current order), the work center is still released. In this case, exceeded means that there is no more capacity available at this work center for the next order. This means that order 107 must be released at work center M. However, work center VK is blocked for other orders.*

*If a subsequent order (order 5 and order 6) loads a blocked work center, it is not released. Therefore, orders 5 and 6 are returned to the MRP inventory.*

## Reorder Point Calculation

The MRP process operates based on known demand, typically referred to as deterministic demand. This demand originates either from customer orders or forecasts.

On the other hand, stochastic demand involves predicting future needs based on past values. In this case, we know the distribution of past demand, which can often be described by a normal distribution. This stochastic demand approach can be used to "manage" a warehouse.

For this purpose, we define the reorder point R. The stock level is continuously reduced by withdrawals. When the inventory level reaches the reorder point R, an order is triggered.

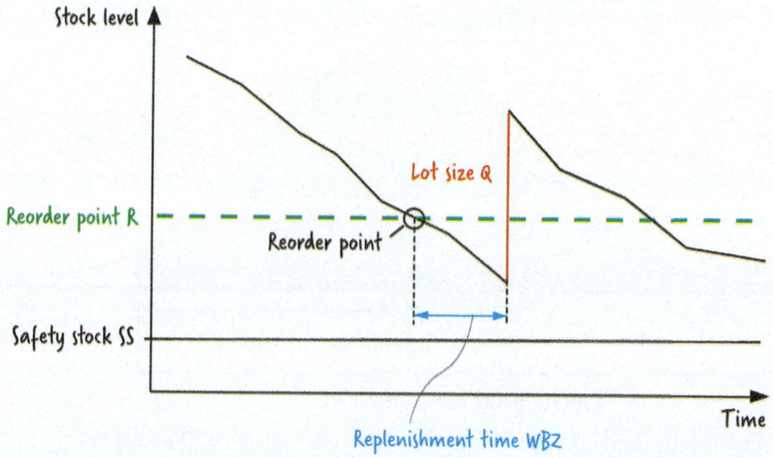


The reorder point calculation is assigned to the methods of the MRP process. For clarity, MRP is the proven integrated approach for managing a warehouse. Understanding this is important because not all MRP processes need to be implemented to perform the reorder point calculation. Therefore, understanding the principles behind this method of the reorder point calculation is valuable. In this way, it can be used efficiently in any production environment and with any type of control system.

**Tip** The reorder point is now calculated from the relationship:

$$Reorder\ Point\ R = Consumption\ over\ Time \times Replenishment\ Time + Safety\ Stock$$

We always assume the worst case: Demand during the replenishment time (WBZ) reduces inventory to exactly the safety stock level.

We further assume that the assumed demand D is calculated from the historical consumption values and that it is normally distributed. This demand is described by the mean consumption μ and the standard deviation $\sigma$.

The safety stock SS is also based on the spread $\sigma$ of the consumption values and is defined as:

Safty stock level $SS = k \times \sigma$.

The factor k determines the service level in a normal distribution—the higher theservice level, the higher the inventory level. However, it also reduces the risk of stock-outs.

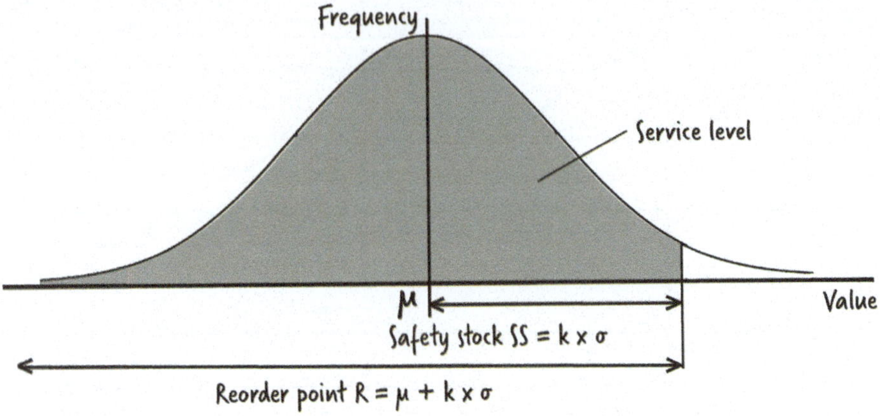


The safety factor k is the numerical value, a multiplier that defines the service level.

| Service level | Safety factor k |
|---|---|
| 50% | 0 |
| 60% | 0.385 |
| 80% | 0.842 |
| 90% | 1.282 |
| 95% | 1.645 |
| 99% | 2.326 |
| 99.5% | 3.09 |

### *Task*

*The plastic housings are manufactured from stock in cost-optimized lot sizes for injection molding. This includes the "pink" housing for the Octopus—a color that is highly requested on specific markets.*

1. *Calculate the optimum lot size $Q_{opt}$ according to Andler. To calculate the Andler lot size, see Sect. 3.3 "Method 11: Single Minute Exchange of Die".*

Weekly Demand D: Mean value 50 pieces/week,
Standard deviation 7 pieces

Lot size calculation: Andler method
Set up cost: 120 Euro
Stocking cost/piece: 0.5 Euro/piece

Desired service level: 95%

Replenishment time WBZ: 2 weeks

2. *What is the right reorder point R to meet the service level? What is the safety stock level?*

***Solution***

1. *The optimal lot size according to Andler is calculated as follows:*

$$Q_{opt} = \sqrt{\frac{2 \times A \times D}{h}} = \sqrt{\frac{2 \times 120€ \times 50 \frac{pieces}{week} \times 52\,weeks}{0.50 \frac{€}{piece}}} = 1118\,pieces$$

2. *The reorder point R is calculated as follows—plus the following hint:*

**Tip** The variance ($\sigma^2$) of the replenishment time WBZ over 2 weeks is equal to 2 × the variance of the replenishment time of one week.

$$\sigma^2_{2weeks} = 2 \times \sigma^2_{1week}$$

Resolved results in:

$$\sigma_{2weeks} = \sqrt{2 \times \sigma^2_{1week}}$$

The generic formula for the replenishment time can be derived from this:

$$\sigma_{WBZ} = \sqrt{WBZ \times \sigma^2_1}$$

with $\sigma$ being the standard deviation of the replenishment time of one-time unit

$$\sigma_{WBZ} = \sigma_{2\,weeks} = \sqrt{WBZ \times \sigma_1^2}$$
$$= \sqrt{2 \times 7\,pieces \times 7\,pieces} = 9.9\,pieces$$

$$D_{week} = 50\frac{pieces}{week}$$

$$R = D \times WBZ + k \times \sigma_{WBZ} = 50\frac{pieces}{week} \times 2\,weeks + 1.645 \times 9.9\,pieces = 117\,pieces$$

An order of 1118 units is triggered when the order stock falls below 117 units. The stock will last for about 1/2 year, which is quite a long time, but normal for a special department.

The Safety Stock SS is: $SS = 1.645 \times 9.9\,pieces = 16\,pieces$

To a certain extent, the reorder point calculation also includes to a certain level the "pull system logic". The stock is not reduced if there is no parts consumption. However, in practice, the reorder point calculation is a push process, as the annual requirement forms the basis for the calculation. This annual requirement is a planned requirement.

## 3.8 Method 15: Line Balancing—Takt Creates Flow

Balancing a production line based on an operator balance chart (OBC) is a common and understandable way for generating flow, as you closely and efficiently try to balance the work content of successive processes or workstations in terms of time, while staying below the required customer cycle. This type of work content balancing was discussed in Sect. 2.6.

Another way of balancing is by order sequence, which is often approached using the EPEI metric. In Japanese, this is also equated with the Lean principle of "Heijunka". However, this is only partially correct, as it would be too short-sighted to limit the heijunka principle to this key figure.

In Japan, Heijunka represents a mindset aimed on reducing "imbalances" or "unevenness" in all their forms. For example, Taiichi Ohno's approach of the Toyota Production System is based on Heijunka as one of the central pillars. This can be addressed through many organizational and technical approaches. For example, "one piece flow" is basically Heijunka, as is the Kanban or the OBC method. The SMED method also reduces batch sizes, making it a key strategyfor achieving balance and equilibrium. In general, any reduction in waste to implement just-in-time move toward Heijunka.

The basic idea of the EPEI approach is to execute a certain number of orders in the same optimal sequence within the shortest possible time frame (e.g., one day). To do this, the EPEI metric is used to calculate the shortest time interval between all orders. The acronym EPEI stands for "every part every interval", i.e., the time between a sequence of orders and the next identical order.

The EPEI is calculated as follows:

$$EPEI = \frac{\sum setup\ time}{available\ working\ time - \sum\left(\frac{customer\ demand\ per\ day \times cycle\ time}{machine\ reliability}\right)}$$

Or to put it another way ...

$$EPEI = \frac{\sum setup\ time}{available\ working\ time - \sum\left(\frac{processing\ time}{machine\ reliability}\right)}$$

The EPEI value therefore indicates whether the orders can be produced in the specified time period (e.g., one day) based on the daily customer requirements. If the EPEI is < 1, then this is possible, meaning, the sum of the setup times of all orders is lesser than the available capacity, reduced by the daily processing times for all customer requirements (considering the respective process reliabilities). In other words, there is enough capacity available to set up all orders for that day. If the value is > 1, the daily customer requirements cannot be met as there is insufficient setup time left. The EPEI metric only shows whether the customer requirements can be processed in the specified time period (in this case, one day) under the boundary conditions of the required setup times, the available working time, the machine/process reliability, and the cycle time. Nothing more, nothing less.

**Tip** The EPEI metric is controversial. Arguably, there are often too many uncertainties and surprises in daily operations. Using customer demand as the central parameter of EPEI is also subject to a high degree of uncertainty. Although this approach is an interesting way of thinking, it would not be a practical planning tool. Rather, the focus should be on "one piece flow" and the ways to achieve it, not on cementing batch sizes but by smoothing them.

We do not want the EPEI indicator to be understood as a scientific and only correct approach to balancing. However, it is an interesting reference point for improvement and discussion. In particular, it is an indicator of whether the customer's requirements can be produced in short intervals without waste. In addition, the discussion of the EPEI often results in an impulse to reduce lot sizes and shorten setups using SMED.

The EPEI value multiplied by the (calculated) daily requirement, and corrected for machine or process reliability gives you an optimal lot size for achieving flow. Based on this, waste can be discussed. What is the current lot size compared to the theoretically calculated lot size based on the EPEI? Can we reduce the number of changeovers? Concurrently, problems with current (process) availability emerge to the surface (see also Sect. 2.4 on OEE).

The lot size calculation according to the EPEI value is then calculated as daily requirement × EPEI.

The Octopus assembly cell currently assembles 5 products. You are wondering if the current assembly logic can be enhanced using the EPEI method to achieve an optimal flow assembly of the different orders. Currently, all products are assembled with a hypothetical daily quantity of 100 pieces per order (current lot size per order). Due to the high demand, the Octopus assembly cell operates in 3 shifts of 8 h each. However, the assembly cell experiences 30 min of idle time per shift. Workers organize all other breaks, so that the assembly process is maintained without further interruptions.

The Octopus assembly cell contains the following requirements and master data:

| Product – final assembly | Daily Demand in pieces | Cycle time in minutes | Processing time in minutes | Set up time in minutes | Reliability in % |
|---|---|---|---|---|---|
| Octopus Standard | 15 | 15 | 225 | 40 | 90% |
| Octopus Small | 20 | 12 | 240 | 40 | 90% |
| Octopus Large | 20 | 11 | 220 | 50 | 90% |
| Octopus USA | 5 | 22 | 110 | 50 | 90% |
| Octopus UK | 10 | 25 | 250 | 50 | 90% |
| Sum | | | 1045 | 230 | |

*Task*

1. *Calculate the EPEI! What does EPEI mean? Start with calculating the processing time considering the reliability factor.*
2. *What can you do to ensure that all products can be manufactured in one day?*

***Solution***

1. *The EPEI is 1.22. In the EPEI calculation, the setup time of 230 min is divided by the delta of the daily working time and the machining time. The effective machining time or processing time for all parts is then calculated as: daily demand x cycle time x reliability, and this time per part is then totaled over all parts.*

| | Working time per shift in hours | Number of Shifts | Daily working time in hours | Daily working time in minutes |
|---|---|---|---|---|
| Daily demand | 7.5 | 3 | 22.5 | 1350 |
| Processing time of the parts (incl. reliability correction) | | | | 1161 |
| Required set up time for all parts | | | | 230 |
| EPEI | | | | 1.22 |

$$\text{EPEI} = \frac{230 \text{ minutes}}{1350 \text{ minutes} - 1161 \text{ minutes}} = 1.22$$

*This means that the daily customer demand for all products, including the yield correction, cannot be produced in one day—3 shift operation, because the setup times excced the available capacity on that day (EPEI > 1). Given the current setup times, the reliability achieved, and the current cycle times, the optimal and shortest sequence and duration is just over one day, if only the lot sizes are changed. However, this is because we have taken the current customer requirements per day as the lot sizes. So, let's examine the lot sizes. As you can see, the lot sizes in the system are currently set to 100.*

| Part | Daily demand | EPEI | Lot size according to EPEI (rounded up) |
| --- | --- | --- | --- |
| Octopus Standard | 15 | 1.22 | 19 |
| Octopus Small | 20 | 1.22 | 25 |
| Octopus Large | 20 | 1.22 | 25 |
| Octopus USA | 5 | 1.22 | 7 |
| Octopus UK | 10 | 1.22 | 13 |

Lot size according to EPEI = Daily demand x EPEI

*Lot sizes for all products using the EPEI method are significantly smaller than the current fixed lot size of 100 units. The formula is based on the fact that with an EPEI of 1, especially the customer requirements for all orders, can be produced per day. If this was the case, the total lead time for all orders would be 1.22 times the daily working time.*

2. *To manufacture all products in one day, setup times must be reduced. The current setup time of 230 min must be reduced by 41 min to 189 min (target: daily working time—processing time of all orders = 0). An improvement in reliability also helps to bridge the capacity gap of 41 min. Due to the rather poor reliability of 90%, 116 min of capacity is wasted (1161 – 1045 min).*

## 3.9 Method 16: Set-Building Simplifies Logistics

By creating sets, you can bundle the supply of parts, thus reducing logistics costs and inventory throughout the process. Instead of supplying individual parts, the supplier delivers a complete set, or warehouse prepares a complete matching set for the assembly workstation. This reduces logistical effort and complexity, allowing the value-adding process to function better. Sets often start with large or expensive parts and are then supplemented by additional materials, such as screws and other small parts. Ikea has perfected the set concept. All the parts needed to build a piece of furniture, including tools, are delivered to the customer. The various parts are then often packaged together as special assembly sets to simplify the assembly process. The whole idea of the set is to simplify and reduce handling.

We will now discuss the advantages and disadvantages of set formation using a practical example.

For a selected variant of the all-rounder Octopus, LeanClean Inc. wants to manage the flexible suction unit, including suction hose and the various plastic suction attachments for floors, furniture, etc. as a set. This set includes 7 plastic parts, a plastic hose, 24 screws, and 5 different rubbers. The idea goes so far that one can even imagine buying the complete set from a supplier and having it delivered. The idea of outsourcing emerged because the plastic hose is manufactured using an extrusion process, and LeanClean Inc. cannot conduct this process in-house.

***Task***

1. *Which topics are relevant for the change? Use the method of the Ishikawa diagram for structuring your thoughts. Please also refer to Sect.* 3.11 *"A3", where the Ishikawa diagram is fundamentally explained.*

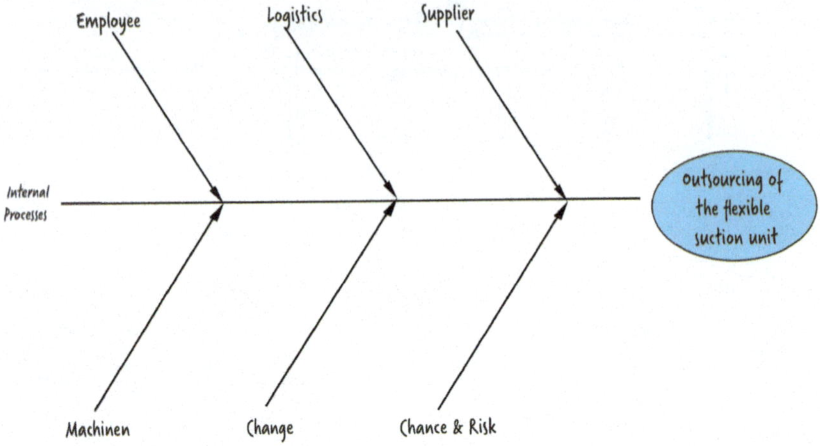


*The Ishikawa diagram above has already very specific, preconfigured branches, and gives you a good start in collecting relevant aspects and issues to consider when managing the transition to the set delivery. Note that the Ishikawa diagram is not used here for the root cause analysis of a problem, but rather as a general cause-and-effect diagram.*

## *Solution*

1. *The following Ishikawa diagram illustrates the various issues you must consider when moving to a set delivery:*

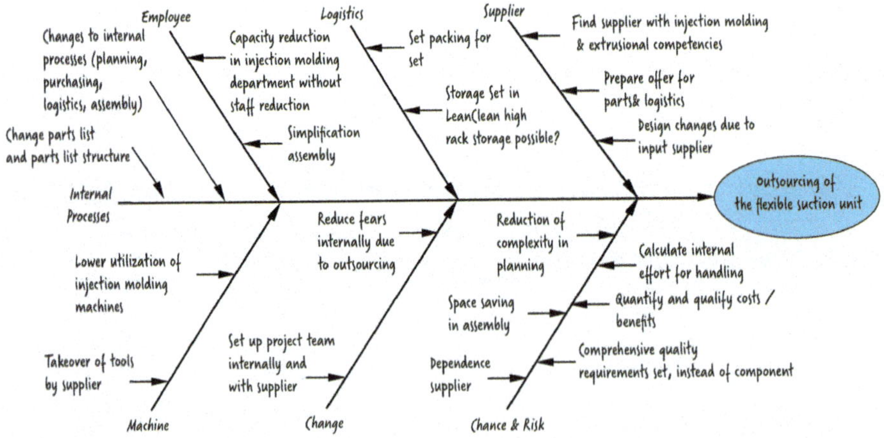

*In addition to the many change-related issues that must be managed, supplier collaboration is certainly of central importance. In addition to the numerous opportunities offered by set formation (complexity reduction, inventory reduction, or space savings), the dependency on the supplier must also be considered in this example. Optimal process integration with a partner is simply not possible without a certain degree of dependency. However, for the otherwise pronounced waste reduction, the negative aspects are less significant.*

## 3.10 Method 17: Sequencing—Simple Sequence Planning Helps Much

Sequencing is an important part of Lean manufacturing. After all, proper sequencing leads to flow and thus reduce waste. Of course, while the sequence is important, sticking to the sequence and not altering it at short notice due to a "boss priority" is crucial. Good planning, transparency, and communication are crucial, due to it's consistency throughout the hierarchy.

Let's examine another specific situation at LeanClean Inc. The large backlog of orders for increasingly complex products is affecting the injection molding department, which is being asked to perform an ever-increasing variety of tasks. For example, many plastic parts (especially vacuum cleaner housings) are now being finished with metal handles immediately after injection molding. This is often a "snap-on" process and does not require any special equipment—a simple assembly process. This process is not performed in assembly but directly in the injection molding department due to its low technical requirements and for flow optimization.

However, the complexity of the injection molding department has increased significantly. In the past, the job was finished after injection molding, but today there are suddenly dependencies between work steps. Until recently, the department's experienced foreman could manually plan operations as needed, but this is no longer feasible due to the process complexity. In particular, a sequence planning for the entire is required.

In injection molding, setup planning is also critical. The molds must be at the machine on time and, most importantly, they must be maintained. This is why transparent process planning, also known as sequencing, is so important. Optimal planning increases transparency, reduces the hectic pace, and minimizes machine downtime. It also ensures better tool maintenance.

Your task is to arrange the orders in a way that minimizes idle times and, in turn, reduces the overall throughput time. The order processing times remain the same. Now, help LeanClean Inc. generate more output by planning early and ensuring visibility.

**Tip** The Johnson algorithm provides a method for determining an optimal schedule. By calculating the optimal order sequence, the algorithm achieves a minimum (total) throughput time.

There are four steps to the process:

(1) Determine the operation with the shortest processing time dij from the processing time matrix D of orders A.
(2) If this operation has the shortest processing time $d_{ij}$ in production process 1, then place this order at the first free position in the order sequence of all orders, in an empty $D_{optimal}$ time matrix. If the shortest machining time dij is for a machining step in production process 2, then place this order in the last possible free position of the orders. This order now has its fixed position in the sequence of all orders.
(3) Then delete this job from the matrix of jobs to be scheduled.
(4) Go back to step 1. Important: All jobs you have planned up to this point remain in their fixed position. The "maneuvering mass of orders" is now reduced by the last planned order.

LeanClean Inc. needs to process the following orders A1, through A6, with the corresponding processing times (in hours) for plastic injection molding P1 and assembly P2, for the next calendar week. An order always consists of operations P1 and P2, which must be processed sequentially. For technological and production reasons, an order always starts with the plastic process (production process P1), followed by the assembly process P2 (see processing time matrix D).

| Processing time matrix D – processing time $d_{ij}$ per order $A_i$ & production process$_j$ | | order | | | | | |
|---|---|---|---|---|---|---|---|
| | | $A_1$ | $A_2$ | $A_3$ | $A_4$ | $A_5$ | $A_6$ |
| Production process | $P_1$ | 15 | 6 | 4 | 13 | 12 | 4 |
| | $P_2$ | 5 | 11 | 6 | 10 | 16 | 2 |

orders per week planned

time in hours

Time for each operation step

$A_1 \ldots A_6$: orders

$P_1 \ldots P_2$: Production processes (different)

$d_{ij}$: Processing time (in hours)

Looking at the original process time matrix, you can see that the different process times of the 6 orders in P1 and P2 can result in unfavorable combinations, leading to waiting and idle times combined with uncontrolled material buffers between process steps. And that, of course, is waste.

***Task***

1. *Transfer the orders from the processing time matrix in their original order A1 to A6 into the template below—this will visualize the initial order schedule. What is the total processing time? At which points in the process are there idle times? Where will material buffers inevitably build up between P1 and P2?*

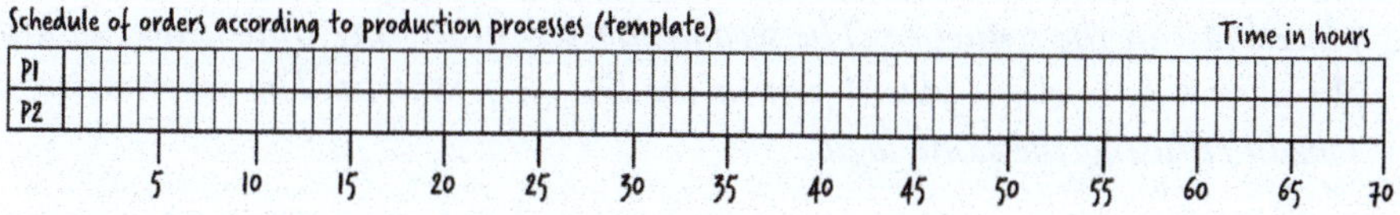


2. *Now determine the optimal sequence for the given orders using the Johnson algorithm based on the processing time matrix D. What is the optimal sequence of orders A?*
3. *Put the orders back into the production schedule for processes 1 and 2. What is the difference in total throughput time? Use the same template again.*

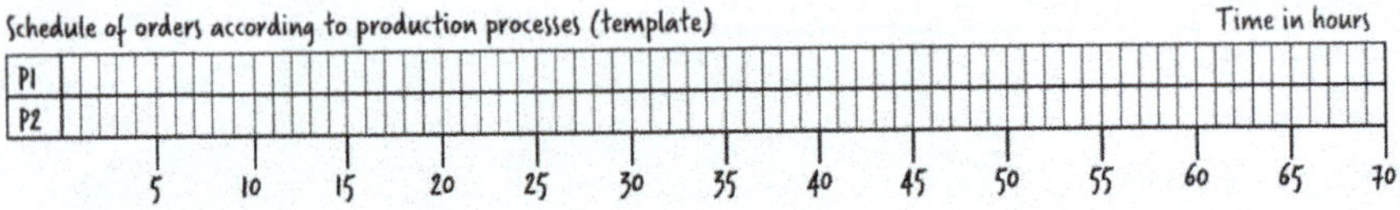


4. *Do you still see wasted time in the production schedule? Please quantify the time losses and compare them to the schedule.*
5. *Is there a way to move orders to reduce or optimize these losses? If not, what else could be done with them?*

## *Solution*

1. Without further optimizations, the schedule would look like this:

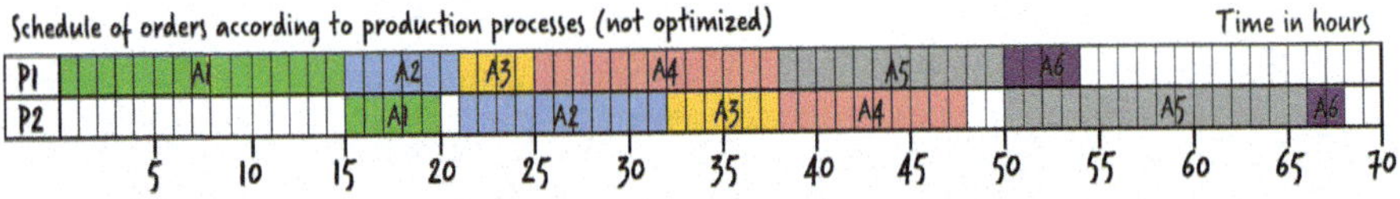


*The total throughput time is 68 h. P2 is not used for 3 h (after A1 and after A4).*

*Although buffers are always created when all parts have undergone a production process, the next production process is not yet ready, resulting in temporary buffer stocks of parts from order A3 before production process 2 and parts from order A6 before production process 2.*

2. *The optimal order sequence is $A_3$, $A_2$, $A_5$, $A_4$, $A_1$, and $A_6$. With 2 h, the absolute shortest processing time in the processing time matrix is $d_{62}$ of order $A_6$. Since this processing time occurs in production process P2, order $A_6$ is placed last in the sequence. The second shortest processing time is $d_{31}$ of order $A_3$ with 4 h, so $A_3$ is placed first in the $D_{optimal}$ matrix. The other orders are scheduled using the same logic.*

Optimal (order) sequence

| Processing time matrix D „optimal" | | Order | | | | | |
|---|---|---|---|---|---|---|---|
| | | A3 | A2 | A5 | A4 | A1 | A6 |
| Production process | P1 | 4 | 6 | 12 | 13 | 15 | 4 |
| | P2 | 6 | 11 | 16 | 10 | 5 | 2 |

time in hours

3. *The sequence from the optimal processing time matrix can now be transferred to the order schedule. In total, by optimizing the sequence, the throughput time is reduced by 11 h from the previous 68 h to 57 h. That's a 16% improvement—and that's just by optimizing the order sequence!*

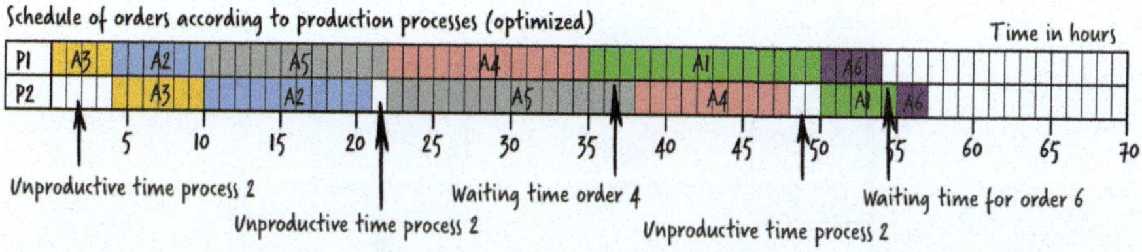


4. *Process 2 has 7 h of idle time, and at the end of the week, the employee in process 1 is allowed to leave three hours early for the weekend. In the initial situation, there are even 18 h of idle time in process 2—not to mention the 14 h at the end of the week in process 1. There are also waiting times: job 4 waits 3 h to be processed, and job 6 waits 1 h to be processed. This is a far cry from the non-optimized state, where a total waiting time of 21 h resulted in significantly higher buffer sizes between process steps.*
5. *If you want to eliminate the waiting times, you can start order $A_3$ in process 2, 3 h later. This gives a time window of 7 h before order $A_3$, but at the expense of newly emerging waiting times. Nevertheless, the employee in process 2 (pure assembly) could uninterruptedly be productive for 7 h. The waiting times of the materials between production process 1 and production process 2 are not so easy to optimize. Reducing the lot size of the assembly processes in order to decrease the processing time, $d_{52}$ and $d_{11}$ is necessary.*

*This shows that all 6 orders could be produced in the injection molding department in 54 h. Given the regular 1-shift operation, overtime must be ordered in $P_1$ and $P_2$, otherwise next week's orders could not be processed.*

## 3.11 Method 18: A3—Managing Improvements Made Easy

### Network Printer Issues

As a special service (e.g., for the elderly), LeanClean Inc. offers to print and send the operating instructions, which are otherwise only available as PDF files, in a particularly legible font size. This offer is taken up to a far greater extent than originally expected by the sales department. As a result, LeanClean purchased three identical laser printers, two of which are located in the sales office while the third printer is located in the hallway outside the office and installed as a network printer so that other departments can access it, too.

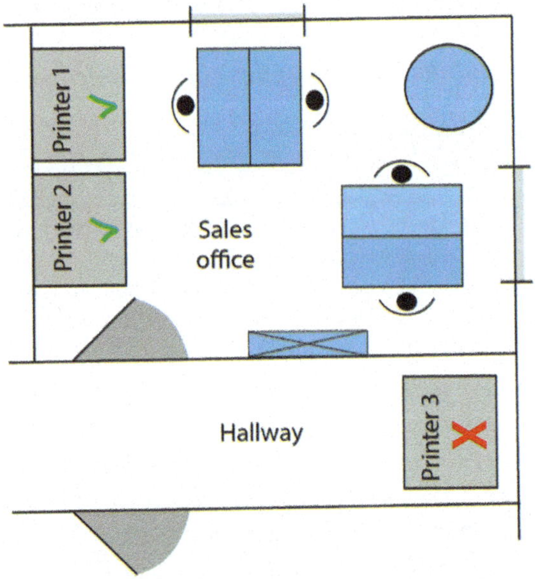


Although all three machines are identical in construction, laser printer number 3 is known for its significant inferior print quality (streaks andpoor color reproduction) and frequent paper jams. Annoyed by this unfortunate state of affairs, the sales manager has an idea: apply the same A3 problem-solving method that successfully solved the quality problems of the Elephant handles.

Of course, you offer your help and follow the steps below:

*Task*

1. *The A3 method works best when it is used by a team that is composed as interdisciplinary as possible. Who would you invite from LeanClean and why?*
2. *Structure an A3 sheet of paper using the template from the textbook "LEAN Production - Simple and Comprehensive". Add a header with the information "Topic", "Solution Team", and "Start Date". Also, fill in the header.*
3. *Describe the "background" briefly and precisely. You will be missing some data in the "problem description"—so draw inspiration from your experience with printers and copying machines. Make your personal assumptions and visualize the problems. Finally, you and your team should formulate the "target state" in relation to the two core problems of "print quality" and "paper jams".*
4. *When analyzing the root causes, you will work with a method known as Ishikawa diagram or fishbone diagram—it is known after its inventoor (you will immediately recognize the similarity). Discuss possible causes—being a "printer service expert", this shouldn't be a problem for you. Give it a go!*

**Tip** First, draw a diagram with up to seven initial branches that lead to the problem description via a horizontal line from left to right. Label the branches with the familiar categories: Man, Machine, Material, Method, and Mother Earth ("5 M"). It is often useful to include two additional categories: Management and Measurement ("7 M").

You can also freely choose the branch's categories. In Sect. 3.9, see the use of the set building methodology. Also, it is important that you always think broadly and comprehensively.

With your team, consider branch by branch, what circumstances, processes, or rules might be responsible for the problem described. This results in sub-branches. In the first round, you can be a bit "wild" when populating the branches; obviously irrelevant causes can simply be deleted later.

Sometimes it is worth pursuing a branch deeper. Don't settle for a superficial answer, and keep asking "Why is that?" until that particular branch has reached the real cause of the problem.

When analyzing causes with an Ishikawa diagram, the space provided on the A3 form quickly becomes too small. Thus, it is best to have a separate template for this.

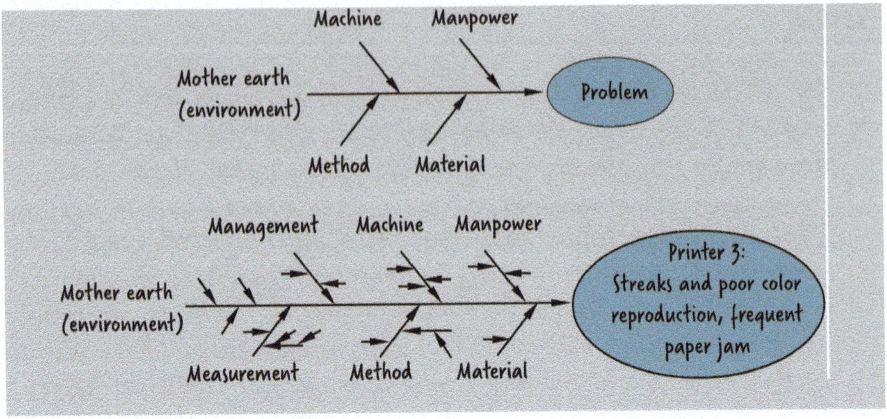


5. *Now is the time to separate the wheat from the chaff. You and your team must prove which of the possible causes of the problem are actually related to the printer's problems. Each branch is, strictly speaking, a causal hypothesis that must be supported by concrete observations, measurements, experiments, or perhaps even calculations. Therefore, carefully discuss how you would proceed in this case to prove the suspected cause-and-effect relationships.*
6. *Against the background of the well-examined—or in this exercise, the most probable—cause-and-effect relationships, you develop concrete countermeasures and finally an action plan with responsibilities and due dates.*
7. *In field 7, enter the metrics or observations you want to use to monitor the effectiveness of the implemented countermeasures.*

**Tip** Go with the problem-solving team to the site of the incident ("gemba"). Get your own perspective on the situation and what's going on.

Even the problem's description should also be quickly quantified. What metrics can you use to prove that the reported problem actually exists?

A thorough root cause analysis is the best foundation for successful countermeasures. Note: Investigating cause-and-effect relationships can take several days or even weeks!

"The more, the better" is the wrong approach regarding countermeasures. Implement only one countermeasure at a time and observe its effect! If you don't make progress, perhaps two actions synergized will induce success. Is your initial assumption about the cause-and-effect relationship confirmed? There is no shame in being wrong—the most important thing is to eventually understand the underlying mechanisms and eliminating the problem for good.

***Solution***

1. *The team should consist of three to five people. The composition should include a: method expert (you), at least one person from sales (problem owner), a (further) person with frequent access to printer 3, a person from purchasing (contact to printer supplier or supplier of IT equipment of LeanClean Inc.), a technical specialist (e.g. printer expert, janitor or a person from maintenance). This first point in the A3 process should not be underestimated. Many projects have failed because the team lacked the "right" people to fully grasp the background and guide the discussion effectively. Having the right team ensures "buy-in" from the most important departments or individuals, which crucial for the subsequent implementation of solutions.*
2. *thru 7, You canmove sequentially through points 2 to 7 of the A3 process (see "LEAN Production - Simple and Comprehensive", section 4.11). You have assembled a good team, with many expertise and experience. The potential causes of the problem are evaluated one by one.*

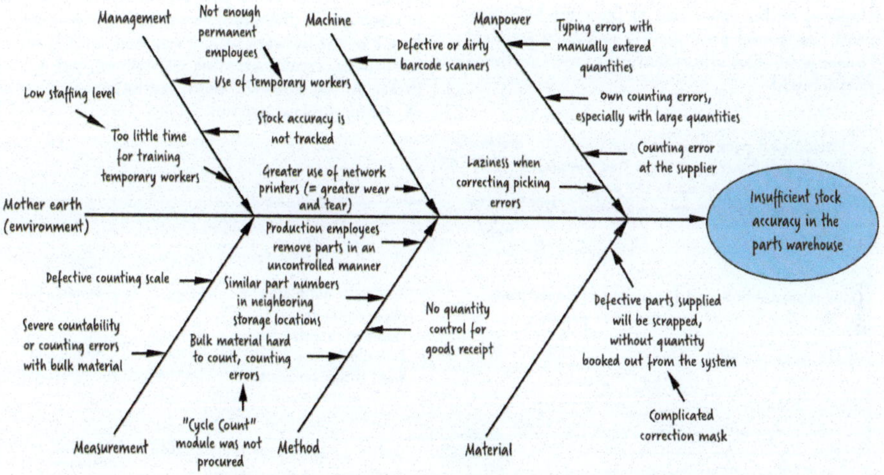


*For example, you could switch to a higher grade of paper for a week and check the quality of the color reproduction and the tendency to paper jams under otherwise identical conditions (print frequency, supply trays used, ventilation behavior in the hallway, temperature, and humidity). If the problem*

*behavior differs, then a cause-and-effect relationship has been identified. As a next isolated experiment, the influence of ventilation behavior could be tested again for another week, using the original paper for printing. The influence of operating errors could be determined by a small survey of users; the printer sensors should be checked for correct functioning by a specialist.*

*By systematically reviewing all cause-and-effect relationships, you have identified the most important improvement measures (step 5 in the A3 process). Ensure focusing on the most relevant actions and formulating them very specifically.*

*But don't get tired yet, as the most important thing is coming: who will do what and by when—without this clarity and agreement, there will be no improvement (step 6 in the A3 process). By "doing", positive results can be achieved quickly. Finally, reviewing and, if necessary, adjusting the countermeasures completes the A3 process (Step 7).*

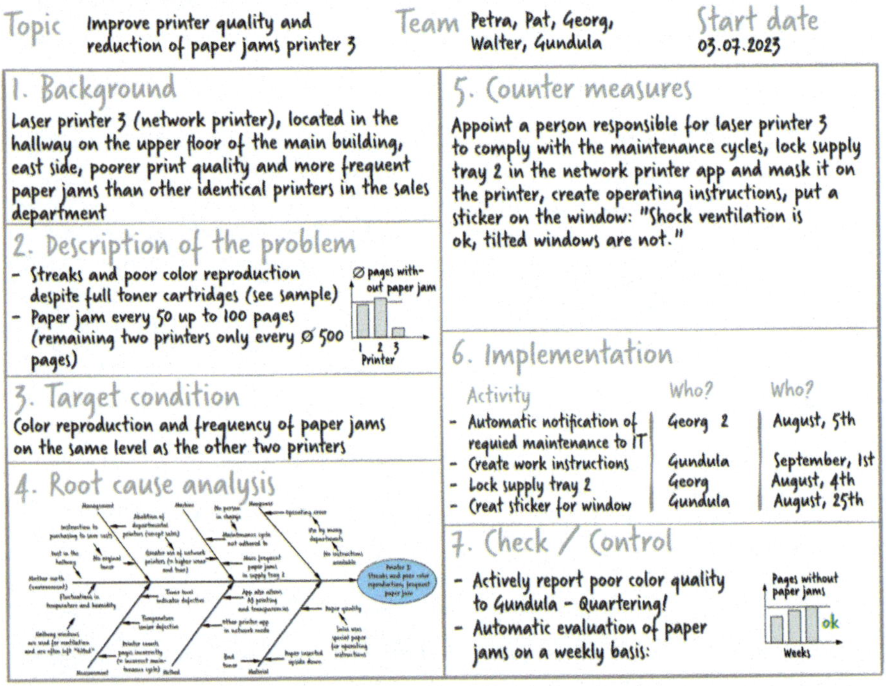


## Inventory Errors in the Warehouse

You met Wolfgang, the goalkeeper's father at your son's soccer practice, and discussed about your work, making Wolfgang impressed with the successful

Lean process at LeanClean Inc. He then mentions that as his employer's warehouse manager and an electric scooter manufacturer, he has been increasingly facing challenges with inventory accuracy, which primarily affects the manual storage area, although there are far fewer problems in the automated small parts warehouse.

You are happy to offer your help—not least, because the two boys get along so well playing soccer. Therefore, the next Wednesday afternoon, you visited Wolfgang in his warehouse and ask him to show you the situation on site. Certainly, after several random checks of some parts, you found discripancies of quantities in the warehouse by up to ten units from what is displayed in the warehouse management system. So, the challenge is real and worthy of the A3 method.

***Task***

1. *Explain who you are putting on the A3 team. Fill in the header of the A3 form.*
2. *Describe the "background", briefly and concisely. When describing and quantifying the "problem", you will be missing some key figures—so use your previous practical experience and other knowledge from various lectures. Formulate the "target state".*
3. *Create an Ishikawa diagram as part of the "root cause analysis".*
4. *Discuss with the team how to investigate and prove the suspected cause-and-effect relationships.*
5. *Now formulate specific "countermeasures" and fill in the "implementation" field.*
6. *What observations and measurements will you use to verify the effectiveness of your actions?*

### *Solution*

*Possible solutions to questions 1 to 4 are available. Appropriate actions can then be derived by examining the cause-and-effect relationships.*

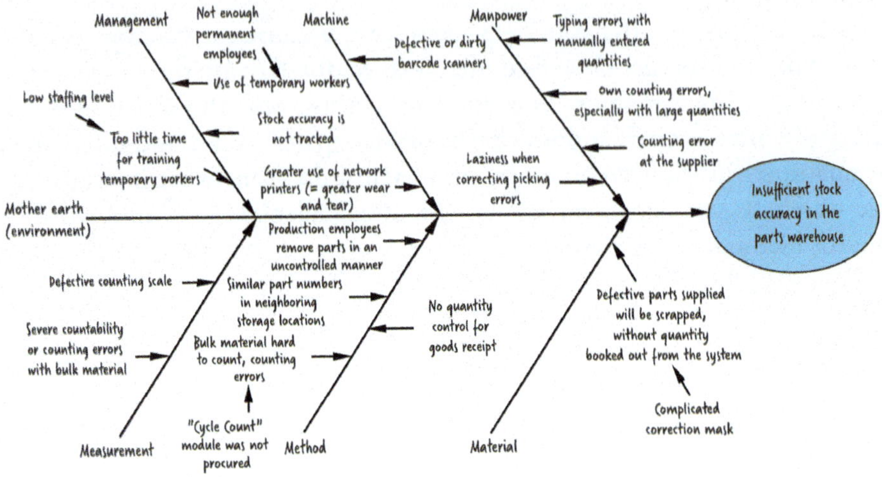


*The Ishikawa diagram is the basis for discussing the countermeasures. The A3 process can then simply be continued from there.*

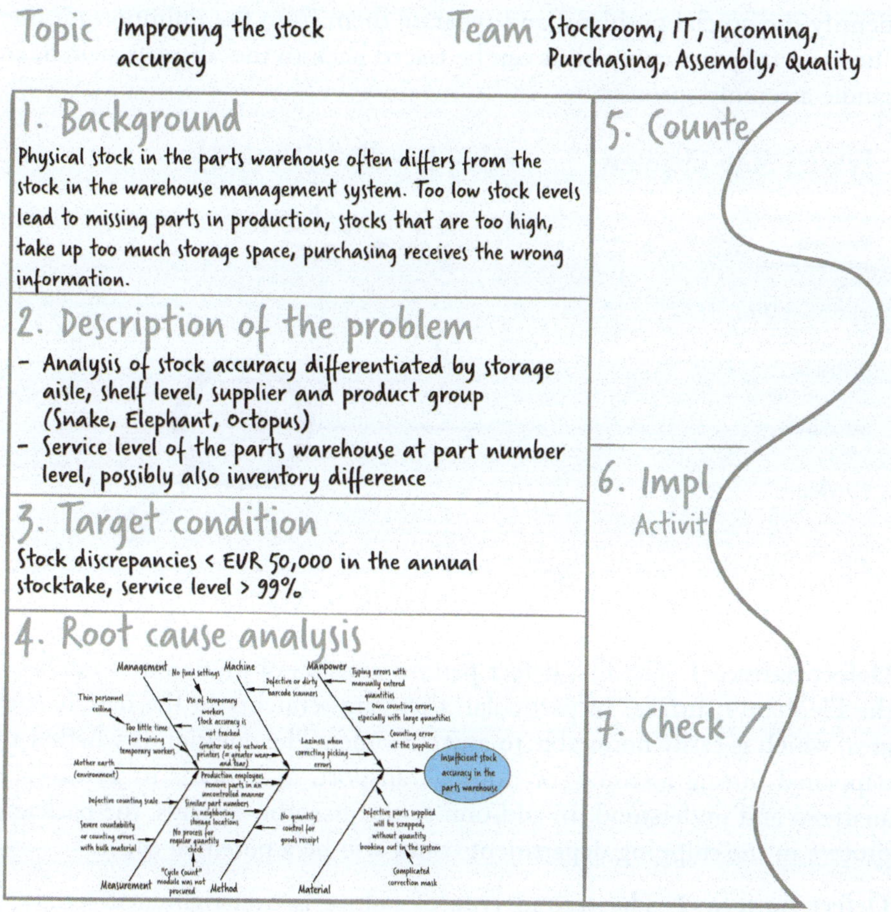


## 3.12 Method 19: Poka-Yoke Prevents Mistakes from Being Made

While the poka-yoke method assumes that people make mistakes, it aims to limit these mistakes through guardrails, curbing their initial occurance or significantly reducing it with a considerable degree of certainty. In common parlance, this is known as "fool-proofing."

The quality assurance, production, and engineering departments are closely examining the various defects in the Elephant, as various defect patterns have been reported based on feedback from the market. Considering being an extremely quality-oriented company, some of the defect patterns are pristine and are of significant concern to LeanClean Inc. The next step is to identify the crucial problems and separate them from the unimportant ones. The most frequent complaints can be traced back to the "display, switch, and handle assembly" process:

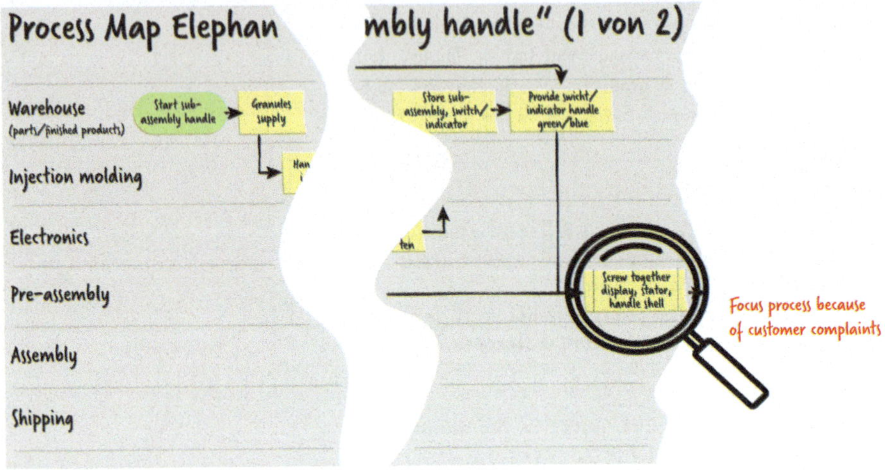


**Defect pattern 1**: The first defect pattern consists of numerous scratches on the Elephant's surface. In particular, there are scratches in the handle shell's area, which is easily noticeable to any customer. The question is whether this happened during assembly or during transport, as it needs to be explicitly analyzed and understood. In addition to the assembly process, the packaging process in the shipping department could also be a potential cause.

**Defect pattern 2**: The second type of failure is over-tightened screws. In addition to the visual "unattractiveness", the question is whether these over-tightened screws can also cause a malfunction. The tightening torque is specified at 30 Nm. However, there have never been any complaints from quality inspection.

**Defect pattern 3**: The third type of defect are wrong screws. Sometimes, the worker is unaware the wrong screws are used during assembly, leading to the switch malfunctions or defects when used by the customer. Investigation of this error pattern revealed that some Elephants are missing screws, which can quickly lead to switch failure and become a safety issue.

## *Task*

1. *Consider other potential mistakes that could occur in the process, given the three defect patterns known from the field. Make feasible assumptions and put them in the prepared table along with the potential root causes described above and apply the FMEA method. What actions should be taken to bring a critical risk priority number (RPN) into an acceptable range?*

**Tip** While the Failure Mode and Effects Analysis (FMEA) is a method for analyzing the risk of possible failures, it is also used for systematic risk analysis of complex products, systems, or processes. The goal of its application is to identify, understand, limit, and eliminate potential design weaknesses and risks to prevent failures in the long term.

The FMEA quantifies the risk of each failure using the risk priority number (RPN). The RPN is the product of three variables: The probability of failure occurrence (A), the probability of failure detection (E) (how can I identify the problem?), and the severity or impact of the failure (S) (often also, the magnitude of the problem when it occurs?). So, the RPN $= A \times E \times S$.

Typically, the individual factors are assigned a value of 1 to 10. If the risk priority number is $> 125$ or an individual factor is $> 8$, specific countermeasures should be developed and implemented. The Ishikawa diagram is often used for root cause analysis.

The FMEA identifies the risks that require action. Corrective actions should be approached using the poka-yoke principle. The simplest and most effective actions will induce sustainable success. The poka-yoke principle is particularly effective in process problems, where several employees are affected by the optimization measures.

In detail, we want to evaluate the three factors: A, E, and S based on the following characterization. It is very important that this assessment is conducted by a (interdisciplinary) team to obtain a comprehensive and coordinated view.

| | | |
|---|---|---|
| 1 | Very low | Probability of occurrence is very low, minor problem |
| 2<br>3 | Low | Probability of occurrence is low, you can deal with the problem |
| 4<br>5<br>6 | Medium | Probability of occurrence is medium, i.e. normal and possible risk, problem has noticeable effects |
| 7<br>8 | High | Probability of occurrence is high, it is very possible that the event will occur, major problem |
| 9<br>10 | Very high | Probability of occurrence is very high, very critical problem |

*Use the following template to structure the issues, their assessment, and possible mitigations—it is intentionally kept simple to speed up the assessment.*

| No. | Error pattern - customer view | Potential cause of error | Effect of the error | A | E | S | RPN | Possible poka-yoke measures |
|---|---|---|---|---|---|---|---|---|
| | | | | | | | | |
| | | | | | | | | |
| | | | | | | | | |
| | | | | | | | | |
| | | | | | | | | |

*Mark all measures with a risk priority number > 125 and discuss the measures using the poka-yoke approach. Are the corrective actions foolproof? What should LeanClean do now?*

## *Solution*

1. *You have now discussed the customer's defect patterns with the team. The quantification of the risk priority number (RPN) is based on the potential causes of defects and the impact of those defects. At this point, it is always a team assessment based on the available information.*

| No. | Error pattern - customer view | Potential cause of error | Effect of the error | A | E | S | RPN | Possible poka-yoke measures |
|---|---|---|---|---|---|---|---|---|
| 1 | Scratches on the surface | Slipping occurs when screwing during assembly | Scratches on the surface of the switch | 6 | 10 | 6 | 360 | Instead of slotted screws, use Torx screws |
| 2 | Scratches on the surface | Packaging causes damage to the surface of the handle scales | Scratches on the surface of the switch | 2 | 10 | 2 | 40 | Change packaging, possibly fill packaging better with polystyrene balls |
| 3 | Screws screwed in too tightly screws | Screws that are screwed in too tightly can damage the switch | Switch failures possible, surface looks "visually" unprofessional | 4 | 6 | 8 | 192 | Use of a torque-controlled screwdriver, documentation of the torque for traceability |
| 4 | Incorrect screws fitted | Installation of incorrect screws | Switch service life limited - vibration on switch increased | 8 | 6 | 3 | 144 | Only one type of screw on the assembly line - set up screws specifically for the assembly line |
| 5 | Incorrect screws fitted | Installation of incorrect screws | Switch service life limited - vibration on switch increased | 8 | 6 | 3 | 144 | For all models, develop handle with only one type of screw |
| 6 | Missing screws | Not all screws fitted | Switch failures in a short time, possible safety problems | 8 | 10 | 9 | 720 | Training employees to use pre-counted screw kits for assembly |
| 7 | Missing screws | Not all screws fitted | Switch failures in a short time, possible safety problems | 8 | 10 | 9 | 720 | Introduce optical Q-system with camera and AI algorithm - documentation for traceability possible |

*It turns out that some potential defects have been classified with a risk priority number > 125. These need to be addressed. You have formulated the actions according to the poka-yoke principle. That's great. A few comments to note: the missing screws seem to be the biggest challenge. This issue only emerged after LeanClean Inc. did a thorough analysis of the customer complaints about the surface scratches. Great—let's work hard to solve this problem. In addition to the necessary training, a "robust" long-term solution is certainly needed. Implementing an optical defect surveillance system with camera and AI algorithm could be the answer. However, this requires investment in IT equipment that needs to be planned and financed.*

*A simpler solution for the missing screws would be to provide pre-counted screw kits. If the worker forgets to put in a screw, the remaining screw in the kit will provide immediate feedback, making it likely the worker will correct the error right away.*

*The engineering department has also started to change the screw type from slotted screws to Torx screws. In the long term, LeanClean wants to reduce the number of screw types and sizes so that mix-ups are eliminated or at least greatly reduced. This will also lead to simplified processes in assembly and upstream production logistics (or at the C-parts supplier).*

*A relatively simple change will be the introduction of torque-controlled screwdrivers. The need for torque-controlled screwdrivers in assembly is currently being assessed. These screwdrivers will then be gradually introduced to all assembly cells and screwdriving processes where torque monitoring is technically necessary. The engineering department is compiling a list for this purpose.*

## 3.13 Method 20: Andon—Transparent Communication

The Andon method is a transparent representation of the current state of production. An Andon board or other information display is designed to show immediately where problems exist. The basic concept of Andon arose from the idea that employees should have the capacity to stop a production line or a selected production process immediately a major issue arises. In this way, the mindset of "empowering people to escalate problems immediately" should become a living culture in the company.

Today, the Andon method is often implemented in production with so-called Andon boards. Central screens, visible from various angles, show managers and employees the current status of production, including any problems. The digitalization process in many industrial companies makes creating Andon boards easier and, above all, the available data is more comprehensive and up-to-date than with the "old" manually maintained boards.

### *Task*

*Now develop a digital Andon board for the injection molding department with 14 machines grouped into three areas. There should be only one board for the entire department.*

1. *First discuss the goals you want to pursue with an Andon board? What is the difference to the shopfloor board?*
2. *Make a concrete proposal for the Andon board in the injection molding department of LeanClean Inc. How would you set it up? What information should it contain?*

### *Solution*

1. *You pursue the following goals with the Andon board; they help to focus on the essentials of the Andon method:*
   - *Create transparency about the situation.*
   - *Provide timely or even real-time information—no delays.*
   - *Make problems visible in a simple and immediate way at the point of occurrence.*
   - *Adequate level of detail for production—not too rough, but detailed enough to suggest immediate actions for improvement.*
   - *Achieve a high level of employee acceptance.*
   - *Define escalation system (when, content, consequences) Shopfloor ⇄ Production Manager ⇄ Operations Manager ⇄ CEO.*
   - *Consistent database for shopfloor board with shopfloor KPIs.*
   - *Automatic creation and updating.*
   - *Flexibility of approach required, dashboard can and should evolve.*

   *The main difference with the shopfloor board is its real-time production status and the simplicity of the information. The goal is to implement immediate measures in the ongoing production. Production workers should do this independently of management. The idea is to "see the problem - act - eliminate the causes sustainably".*

   *The purpose of the shopfloor meeting is to reflect with all employees on the previous day or the current situation and to plan for the near future. To this end, measures are defined that are also effective in the short term, but not immediately. It is often necessary to tackle improvements that require a certain amount of planning and implementation time. However, these improvements should have a short-term effect - in the sense of a target period of "days" rather than "weeks".*

2. *The following Andon board is one way of implementation:*

   *It makes sense to create a separate Andon board for each area, as one board with 14 machines is rather confusing. And the question is whether all employees would have an unobstructed view on such a large Andon board. Therefore, dividing it into the different injection molding areas is sensible.*

   *The board below shows a possible implementation for an area consisting of three plastic injection molding machines, including dependent processes such as assembly robots or packaging. After much discussion, you have decided to use OEE as the relevant KPI.*

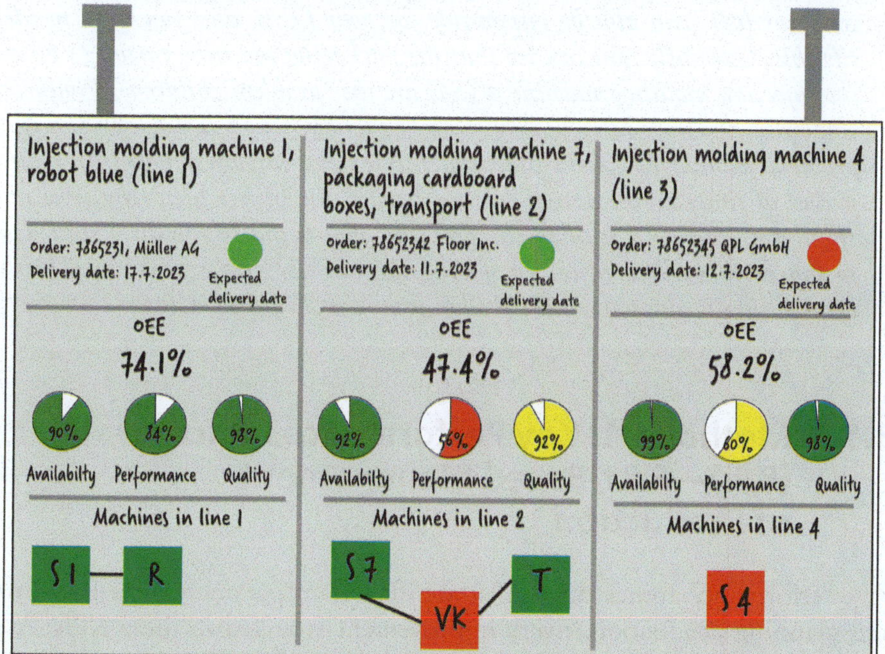


*The Andon board displays 3 data sections for each machine. The first section shows the current order with customer name and order number. In addition, the expected on-time delivery for this order is estimated. LeanClean has solved this on-time estimation in a simple way: it compares the current output of the line in units with the target output. To do this, it performs a linear extrapolation of the actual output over the throughput time and then compares this result to the delivery date. This serves only as an indication of possible schedule problems—nothing more, nothing less.*

*The second section is the OEE metric of the line, which shows where the problems currently lie. Please refer to Sect. 2.4 for more information on OEE. LeanClean Inc. has decided that this metric is crucial for the profitability of this department; it even serves as central KPI for the entire company. A distinction is made between the degree of utilization, the degree of performance, and the degree of quality, and it is these three values that are presented. Management has defined what is "green", "yellow", and "red" based on the target system and the budget.*

*While the lowest area shows the respective problem machine that might induce a poor OEE value, this is essential, as the OEE of a line depends on each individual machine. Unfortunately, it is often not the most expensive and complex machine that is decisive for a good OEE, sometimes a "small conveyor belt" can also be responsible for poor OEE, and you have to identify this. Currently, you can see that the packaging machine carton (VK) and the injection molding machine 4 (S4) are the "problem children". Production employees can now immediately recognize where action is needed. And if the problems cannot be solved independently by the employees within a defined period of time, an escalation cascade toward the higher hierarchy level takes effect; this can even lead to the problem area being temporarily shut down to concentrate all resources on solving the cause(s) of the problem. This last consequence is also part of the Andon process at LeanClean Inc.*

## 3.14 Method 21: Key Performance Indicator (KPI)—Fact-Based Management of Production

Key Performance Indicators (KPIs) are important parameters for measuring and planning production. Every improvement is linked to these KPIs. After all, you must be knowledgable about production. In addition to the current status, these KPIs also show potential waste. This transparency helps to develop a common understanding of the current situation and demonstrate the need for improvement to all employees and management. According to the management principle "what gets measured gets done", we will closely examine some of the KPIs. For more information, please refer to section 4.14 in the book "LEAN Production - Simple and Comprehensive".

Production revolves around three strategic priorities: "deadlines or time", "cost", and "quality". Of course, there are many other relevant issues in a company that can be measured by KPIs. At LeanClean Inc., production is focused primarily (if not almost exclusively) on these three priorities. Let's explicitly discuss them:

***Tasks***

1. *Scrap and yield: In production order A-234512, the injection molding machine produces 1000 plastic housings for the elephant, 20 of which were rejected due to quality problems. Calculate the scrap rate and the yield. Yield is the percentage of good parts or the yield of a process (before rework).*
2. *Degree of efficiency (based on units): 20 h (2.5 shifts) were worked at an assembly workstation to assemble the "handle" assembly. However, the production order only requires 1000 assemblies during this time. However, 1200 units were produced that day because there was too much material delivered from the warehouse to the assembly work center. What is the efficiency of the work center on this day? Discuss the advantages and disadvantages of overproducing.*
3. *Efficiency (based on hours): The work plan for the CNC-turned shaft of the Elephant's handle indicates a time requirement of 57 h. However, the employee reports 84 h for this job. What is the efficiency for this job?*
4. *Capacity utilization: LeanClean Inc. produces 12,000 elephant handles this year. However, the pre-assembly cell could produce 18,000 elephant handles*

*per year with 4 people working 2 shifts. What is the capacity utilization for this year?*

5. *Productivity I: The total labor productivity of employees is calculated by dividing the goods produced by the total number of hours worked by a company's employees in a given period. Suppose a manager wants to calculate the productivity of all employees in his production. The manager calculates that the company produced 30,000 units last month, while the input was 3000 man-hours. What is the labor productivity?*
6. *Productivity II: The CEO of LeanClean Inc. wants to calculate how productive the pre-assembly department was last year. To do this, he calculated all the actual attendance hours of the employees, which totaled 31,000 h. The gross hours (after deducting all vacation time) would have been 33,000 h—and since LeanClean Inc. pays its industrial workers time-and-a-half, they are paid the same amount each month. The reported hours for all orders totaled 22,000 h. He now puts these figures in relation to each other and determines first the attendance value, and then the attendance productivity. What values did the pre-assembly cell achieve? Discuss the results—what can be done to improve the productivity?*
7. *Productivity III: Another common way to measure a company's productivity level is to divide total sales by the total number of hours worked. For example, LeanClean Inc. has net sales of €30 million and its employees worked a total of 95,000 h in the last fiscal year. The output is the company's net sales and the input is the number of hours worked. Calculate LeanClean Inc.'s productivity. Is it a good value?*
8. *Production time and lead time: A production order for 2400 gears consists of the following times: Setup time 4 h and machining time (piece time) of one gear 45 s. The various idle times and administrative times (transport, buffer stock, machine downtime...) total 67 h for the order in question. What is the total lead time of the order? What is the actual total production time of the order? What is the difference between the "lead time" and the "production time" of an order?*
9. *Delivery capability: The current stock of the Elephant is 987 pieces. In the next 4 weeks, another 2400 Elephants will be added from production. An important customer "Retail Best Price" orders 3750 units with a delivery date in 4 weeks. What is the delivery capability in %?*
10. *Value-added time: The pure production time of an order is 500 min. The administrative time to create the order is 50 min. The idle time in production is usually 3 times the actual production time. What is the "value-added time" of the order in relation to the total throughput time? We use the term "value-added time" to refer to the time in which actual value is added to the product.*

11. *Cashflow Management: Today, cash flow management is more important than ever. This KPI is particularly relevant in times of increasing interest rates, such as when a company faces higher inventory cost due to supply chain issues. Managing the cash-to-cash cycle effectively is therefore important. For the quarter ended September 30, 2008, LeanClean Inc. had a cash-to-cash cycle consisting of the following values:*

    - *Average debtor payments: 36 days*
    - *Average storage time: 8 days*
    - *Average vendor payments: 69 days*

    *What is LeanClean's cash-to-cash cycle? What does this number really tell us? Explain the result.*
12. *Try to structure the various KPIs from the previous 11 exercises into the categories of "time", "quality", and "financial". Add more KPIs to the above list—brainstorm in a team. The KPIs should relate to an industrial company and cover all operational tasks in production. Prepare a one-page summary of your discussion.*
13. *The various KPIs should always be presented graphically—a diagram often says more than a number. You can also see the evolution of the KPI over time. Let's take the KPI "production lead time. The two products "Octopus Pink" and "Elephant Blue" have the following data:*

| Week | Production lead time "Octopus Pink" (days) | Production lead time "Elephant Blue" (days) |
|---|---|---|
| Week 11 | 23 | 28 |
| Week 12 | 21 | 27 |
| Week 13 | 23 | 25 |
| Week 14 | 35 | 25 |
| Week 15 | 26 | 19 |
| Week 16 | 28 | 23 |
| Week 17 | 21 | 22 |
| Week 18 | 34 | 21 |
| Week 19 | 32 | 25 |
| Week 20 | 34 | 22 |
| Week 21 | 35 | 28 |
| Week 22 | 32 | 16 |
| Week 23 | 39 | 15 |
| Week 24 | 38 | 16 |
| Week 25 | 34 | 19 |
| Week 26 | 29 | 21 |
| Week 27 | 29 | 23 |
| Week 28 | 24 | 19 |
| Week 29 | 36 | 23 |

(continued)

(continued)

| Week | Production lead time "Octopus Pink" (days) | Production lead time "Elephant Blue" (days) |
|---|---|---|
| Week 30 | 34 | 21 |
| Week 31 | 34 | 17 |

*Use this data to create a KPI chart. Also think about how you can turn the data into information that induces change—make reasonable assumptions to realize this. What do you need to think about? What is important additional information that a "reader" of your KPI chart needs to know?*

## Solution

1. $Scrap\ factor = \frac{number\ of\ rejected\ parts}{total\ number\ of\ parts} = \frac{20\ pieces}{1000\ pieces} = 2\%$
   *Regarding the term "yield", and also for the term "first pass yield", revisit the task in Sect. 2.8 "Calculating First Pass Yield and Yield". In the following, we do not distinguish between yield and first pass yield. We are only calculating the yield.*

2. $Efficiency\ factor = \frac{1200\ pieces}{1000\ pieces} = 120\%$
   *An efficiency of 120% is not sustainable over a long period of time, as capacity should not be constantly above 100%. In addition, no production should be accurate (e.g., waste, overproduction, inventory...) or necessary to meet customer demands.*

3. $Efficency = \frac{57\ hours}{84\ hours} = 67.8\%$

4. $Capacity\ utilization = \frac{12{,}000\ pieces}{18{,}000\ pieces} = 66.7\%$

5. $Manpower\ productivity = \frac{30{,}000\ pieces}{3000\ hours} = 10 \frac{pieces}{hour}$

6. $Attendance\ value = \frac{31{,}000\ hours}{33{,}000\ hours} = 94\%$; $Productivity = \frac{22{,}000\ hours}{31{,}000\ hours} = 71\%$
   *In total, the pre-assembly department at LeanClean Inc. realized 33% waste, which means that no value was added in 33% of the paid labor hours (22,000 h out of 33,000 h).*

*Actions to optimize attendance: culture, teamwork, training…*
*Actions to optimize pre-assembly productivity: many lean optimizations.*

7. $Turnover\ productivity = \frac{30\ Mio.€}{95{,}000\ working\ hours} = 315 \frac{€}{working\ hour}$
*It is impossible to ascertain if it's a good or bad value. Productivity depends on the actual value added per hour, not on the sales per hour. It can only be an indicator. If €270 of the €315 in sales per hour are purchased parts, the value added by LeanClean is only €45 per hour. At an hourly rate of €60, this would be a poor value.*

8. *The actual total production time of the job is:*

$$Production\ time = 4\ hours + \left(45 \frac{seconds}{piece} \times 2400\ pieces\right)$$
$$= 34\ hours$$

*This can also be used to calculate the operating time (this is a theoretical value per part, e.g., for cost accounting purposes).*

$$Operating\ time = \frac{34\ hours \times 3600\ seconds}{2400\ pieces} = 51 \frac{seconds}{piece}$$

*It is important to note that the setup time is added to the time per piece in the operating time. Operating time is different from time per piece and setup time.*
*The throughput time is calculated by adding the production time to all waiting times (administration time, idle time of the products in production or in a warehouse…). This gives a lead time of 101 h (67 h + 34 h).*

9. *Available parts for the next 4 weeks = 987 pieces + 2400 pieces = 3387 pieces. This is insufficient to meet the customer demands.*
*The delivery performance is* $\frac{3387\ pieces}{3750\ pieces} = 90\%$. *This is not a good value for a production in a high wage country (such as Switzerland or Germany).*

10. *Value-added time in % of the total throughput time:*

$$\frac{minutes}{50\,minutes + 500\,minutes + 1500\,minutes} = 24\%$$

11. *Cash-to-Cash-cycle = accounts receivable time + inventory time - accounts payable time = − 25 days. This means that the suppliers are financing LeanClean Inc. because LeanClean Inc. can collect customer receivables faster than it can pay creditors. In addition, the delta between accounts receivable and accounts payable is still sufficient to finance the inventory.*

12. *List of KPIs by category:*

| *Time* | *Quality* | *Finances* |
|---|---|---|
| *(Capacity utilization)* | *Scrap* | *Efficiency factor* |
| *Production time* | *Yield* | *Efficiency* |
| *Lead time* | | *Capacity utilization* |
| *Delivery performance* | | *Productivity* |
| *Value-added time* | | *Cash-to-Cash-Cycle* |
| *Idle times* | | |

*It is not always clear what a particular KPI should be classified as. This is not critical because what is important is that you are aware of it. Also, this will impact how you use the KPI as a management tool and where you can look for improvements through the transparency of the KPI.*

13. *A KPI chart has several important elements and aspects that need to be considered for a KPI to be "lived" and have an impact. The chart must not leave any questions unanswered, otherwise the data and the presented facts, e.g., for the shopfloor, will not have any impact. A KPI chart is also much more than a simple graphical evaluation of numbers.*

- *Define target audience—"Who was the KPI chart designed for?"*
- *Title—concise name*
- *Axis labels including measurement units*
- *Trend lines*
- *Definition of KPIs, including data fields and any formulas used—determination of KPI update frequency*
- *Storage of historical KPI data*
- *Creator and creation date*
- *Communication channel*

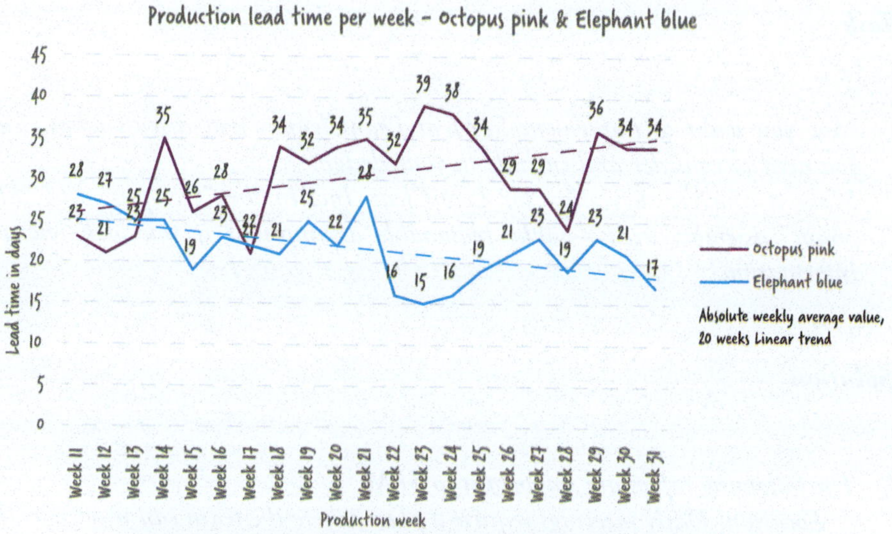


Definition: Production lead time / data: ERP data from MyPlan, production feedback - production start + 1 day, weekly evaluation with average values of all orders

Author: Logistics / to: all department heads, each Monday 11:00 by email

Communication to employees via shopfloor meetings in production, logistics, PPS

Fk, 9.8.

## 3.15 Methode 22: Shopfloor Management with the Involvement of the Entire Team

You have explicitly covered the topic of KPIs. But how do you ensure that these metrics drive action? One promising approach is Shopfloor Management (SFM). We at LeanClean Inc. would like to introduce shopfloor management. In SFM, information relevant to management is gathered on a departmental basis, the degree of target achievement is discussed, and specific actions are derived in the event of deviations. SFM meetings should occur daily, be brief (15 minutes or less), and be held throughout the company. Participants in a SFM meeting come from the respective department, but also from neighboring departments with which there is an internal customer–supplier relationship. Ideally, the meetings occur directly at the place of action ("gemba," you surely remember)—i.e., in production or in the warehouse, and in sales, as it is much better than in meeting rooms.

### *Task*

1. *Now you want to implement SFM at LeanClean. First, discuss what issues you need to consider and include in your design.*
2. *Make a concrete suggestion for LeanClean Inc. in the "Octopus" and "Elephant" assembly areas—make reasonable assumptions if you lack relevant information.*

### *Solution*

1. *The following topics are important to SFM:*

   *Location: Daily meetings require a place where the team can meet. For this purpose, it is important to have a blackboard or whiteboard and a flipchart to document actions.*

   *While this seems simple, it can be difficult to solve in practice, especially in production areas. The shopfloor board must be clearly visible and not "obstructed" by production material. There must also be enough space for the meeting participants.*

   *Rules: A few key simple key rules must be followed for the daily shopfloor meeting. Keeping the topic concise while avoiding long explanations. Avoiding a discussion of the solution is useless at this point, as this should be done after the meeting. Focus only on deviations, as there's no need to dwell on what's working well. Punctuality is critical, and a deputy should always be assigned to lead the meeting if needed. It must also be clear who from each department will attend.*

   *Key figures: Key figures are needed to show current issues and progress. However, it's important to keep the number of KPIs limited to what's relevant for the shopfloor. These KPIs can and should change. For example, new knowledge about current issues may emerge over time, necessitating adjustments to the KPIs. The KPI should be simply presented with sufficient detail and content. The charts should not be complicated - simplicity wins and creates understanding.*

   *Cascading: The shopfloor culture should be embedded throughout the organization. It is therefore important that all areas (including indirect areas) implement shopfloor management. It is also important that there is meaningful cascading in the company—from the individual production area to the shopfloor at management level. For an SME with approx. 300 employees, 3 to 4 management levels are appropriate.*

*Buy-in: Employee buy-in is very important to the shopfloor process as it is significant regarding transparency and participation. However, this approach may be new to many employees and is not always desired. During a shopfloor meeting, all the problems and challenges in the departments become apparent as everything is on the table! This culture of "openness" and "actuality" creates many benefits but requires much time and many discussions with employees.*

*Digitalization: Recently, there are many discussions in the industry about the "digital shopfloor". Although digitization KPIs is a good thing that brings many benefits, it also has its limitations regarding "flexibility", "employee buy-in", "simplicity", and "cost". Our suggestion: start "non-digital" and optimize the shopfloor sequentially by using digital options.*

*To-Do: The heart of shopfloor management is the to-do list. Since discussion leads to action, documenting these actions immediately and transparently during the shopfloor meeting, while including responsibilities and deadlines is paramount. In addition, the short daily status meetings on the actions are crucial for the success of the shopfloor process.*

*In numerous cases, companies often have processes in place to organize and monitor improvement activities or to initiate new activities according to the CIP methodology. The shopfloor measures should be considered as "just do it" measures, meaning they must be implemented directly by the shopfloor area. If complex or long-term improvement measures emerge from the shopfloor, they must be transferred to the existing CIP processes for follow-up; their follow-up is then no longer the task of the shopfloor.*

2. *Now compare your solution to the approach taken by LeanClean Inc. The shopfloor for the "Octopus" and "Elephant" assembly was organized as followed:*

   *Location: The location of the shopfloor board in the production area is very close to the entrance. The space for the whiteboard has been determined. The whiteboard has been given the title "Shopfloor Assembly Octopus & Elephant" and the floor has been marked with yellow tape so that the space is not "accidentally" used as a temporary storage area.*

*Rules: A few simple rules have been established for the shopfloor.*

- *Time: Daily 8:00–8:15 a.m. (1st shift), 3:00–3:15 p.m. (2nd shift)*
- *Attendees: All production employees, maintenance managers, schedulers*
- *Additionally on Wednesdays: Participation of quality an industrial engineering*
- *Chairperson: Fritz Z., Deputy: Klaus K.*
- *Short speeches! Focus on "doing" only*
- *Procedure: KPIs in four categories—Comments—"To-Do's".*

*Key figures: LeanClean Inc. took existing KPIs for on-time delivery, cost, and quality and adapted them slightly for each department. Although not completely defining new KPIs during shopfloor implementation was important, some of the KPIs had to be modified and simplified, as they are constantly evolving to reflect what is relevant today. Considering that the goal is to initiate improvements, and not to provide general information, only the information of the respective department is presented. The assembly departments now measure order on time, next week's orders, target vs. actual hours, productivity, scrap, and top Q issues for the week.*

*Assembly has also developed new information in the area of "people". For example, it is important to have transparent scheduling, including vacations and absenteeism, on the shopfloor board, while the employee skills matrix serves as an important tool for improving departmental flexibility.*

*Cascading: While cascading is essential, LeanClean Inc. has established three of the shopfloor meeting levels: Production areas (Level 1), the entire Operations area (Level 2), and Executive Management (Level 3, EM). All departments (e.g., Sales) define if they have Level 1 and Level 2 for their area. It is important that the designated department delegates, who have already participated in their Shopfloors, meet daily at 11:30–11:45 for the EM shopfloor. The whiteboard is hung on the wall outside the large meeting room—a central location with enough space to meet briefly.*

*Buy-in: The buy-in of the production workers at LeanClean has been impeccable. The quality and relevance of information for each production employee has increased significantly, thereby immediately helping everyone in their daily work. While it was more difficult in the other departments, there were often comments such as "Are you trying to control me now?", "I don't have time for this", or "This is for production, not finance". Fortunately, these negative attitudes diminished over time as the benefits became tangible to everyone.*

*It significantly helped that the CEO of LeanClean Inc. is convinced of the benefits of SFM and sets the "pace" through the management shopfloor ("leading by example", "walk the talk"). SFM can only work if top management stands 100% behind it as leadership is important. After all, the entire management process at LeanClean Inc. is being changed.*

*Digitalization: While different departments were collecting many metrics, each department was organizing them separately. When implementing shopfloor management, LeanClean opted infor a special location in the LeanClean cloud where all the data and charts are stored. Each department manager can access this, create their charts or use the existing ones. There was much discussion about the "inefficiency" of this approach, as many "bosses" now have to create charts. However, the CEO is convinced that it is important for line managers to create (good!) charts. That way, they always deal with the facts immediately. While a more efficient system may be adopted later, this approach is effective for now.*

*To-Do: The successful implementation and tracking of shopfloor actions significantly impacts the operational improvement process. LeanClean has simplified this by hunging a flipchart next to the Shopfloor board, for noting actions and their status. Major issues are entered into the central CIP tool, where they are prioritized and tracked. The focus in the departments is clearly on "doing".*

*After a short time, LeanClean Inc. noticed a significant reduction in the number of fire-fighting meetings throughout the day. Many issues are discussed briefly during the shopfloor and actions are defined. Cascading has also resulted in topics being addressed "hierarchy-specific", but also exchanged across hierarchies in a very structured way. Shopfloor management leads to fact-based, more transparent, and much faster communication.*

## 3.16 Method 23: Change—How to Manage the Implementation of Lean?

As part of your studies, you have encountered several concepts for driving innovation and managing large-scale projects in companies. In particular, you remember the Lean lecture and chapter 5, "Lean Change", from the textbook "LEAN Production - Simple and Comprehensive".

Lean transformation only works when "wanting" and "knowing" form the basis of a comprehensive vision of the future state. You also understand that a comprehensive Lean vision enables good planning, especially of bottleneck resources. This greatly simplifies the implementation of a far-reaching change project.

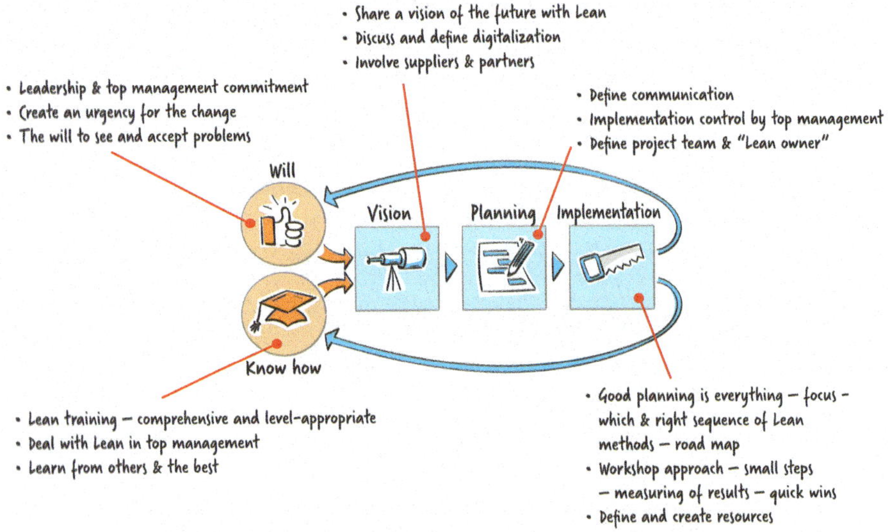


Each element of Lean change has different aspects to consider.

**Tip** The initiative, the "will", must come from the top. Never start without this clear commitment. It must also be clear to all employees why the change is necessary. Top management must face the problems openly and address them without seeking to place blame. That's how you improve.

Knowledge is also key—comprehensive training at all levels is a must, and top management must be actively involved in Lean. This can be supplemented with visits to other companies, not to copy them, but to learn and understand how Lean transformation succeeds.

Developing a Lean vision, with input from suppliers and partners, leads to what is known as North Star of change. The opportunities and risks of digitalization must also be considered. Digitalization is an opportunity for Lean—but it is always "Lean first".

How Lean changes are communicated is vital for gaining broad acceptance and encouraging active participation. This leads to many valuable ideas from the grassroots, which are necessary for successful change. It is also important to appoint a Lean champion for the project, supported by a dedicated team. Of course, top management must take responsibility and review the progress of the change, through regular steering committee meetings.

Results only come from action, but nothing works without proper planning. Detailed planning is crucial for maintaining focus, as it doesn't hinder creativity but provides stability. Many approaches such as workshops, small steps, and clearly visible quick wins in the form of pilot projects should be included in the plan. Resources are also essential—they must be identified and freed up from other tasks.

You will also recall from the "Change Management" lecture, there is a framework that suggests eight sequential stages for a successful, strategically designed change process. This model details the three steps and two prerequisites from the figure above, and is briefly presented below.

**Eight-step process for successful change**

The eight sequential stages in the figure below were developed by Harvard professor John P. Kotter after having studied hundreds of change processes in companies. The lessons learned, especially from major mistakes, result in a logical sequence of prerequisites and actions that should be applied sequentially.

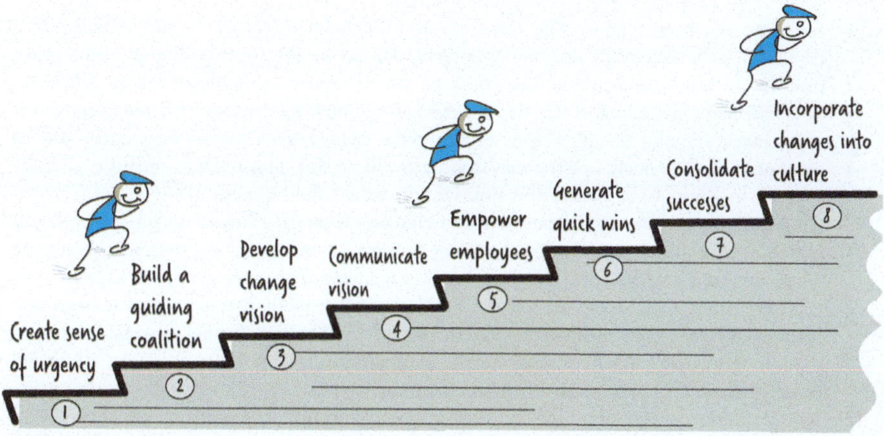


**Tip**

(1) Create a sense of urgency: For change to happen, a critical mass within the organization must not only understand the need and urgency for (behavioral) change but also feels it. Overcoming human inertia and complacency is important. Visible crises and negative news about dramatic market losses and dissatisfied customers is significantly

impactful, likewise new opportunities, such as new technologies and business processes, or particularly challenging goals that cannot be achieved with normal efforts.

(2) Build a guiding coalition: A major transformation can't be led and controlled by one person. Depending on the organization's size, an inner leadership circle, from half a dozen to several dozen people gives breadth and depth to the initiative. The criteria for selecting these team members are: hierarchical importance, expertise, credibility, and leadership skills. Also, all members of the leadership coalition should share a deep sense of urgency for the change at hand and are perceived as the company's opinion leaders.

(3) Develop a change vision: Everyone in the company deserves to know what the future will look like, why you want (or need) to get there, and how you will get there. The leadership coalition should develop these answers and translate them into a powerful vision that can be coherently communicated in no more than five minutes (1-min executive summary) and that speaks to both the heart and the mind.

(4) Communicate the vision for buy-in: To be effective, the vision must be communicated broadly, using all opportunities and media. You cannot over-communicate. The most effective form of communication is a credible demonstration of new values and behaviors by top management and members of the leadership coalition. This may sound a lot like one-way communication, but it also involves listening!

(5) Empower broad-based action: Despite good intentions, many employees face barriers from taking actions positively align with the company's vision. Typical barriers are often found in organizational structures and existing rules. However, people need to unlearn old behaviors and practice new ones, which requires time, patience, education, and training. In other cases, some managers may be "at odds" with the new concepts and therefore need to be removed from their current position.

(6) Generate short-term wins: The best and most convincing proof of the viability of a new idea is its successful and rapid implementation in the form of highly visible pilot projects. In Lean, for example, this could be the perfect introduction of Pull and Kanban in a selected operational area. By impressively comparing metrics such as throughput time, inventory, and quality before and after the change, even the harshest critics can be convinced of the benefits of the new approach. Successful pilot projects must be carefully selected in advance, to ensure and positive impact on the business.

(7) Consolidate success and initiate further change: Now that most of the company is hopefully convinced of the new concept's benefits, reducing its commitment based on what has been achieved will be fatal for management. The opposite must happen: The achieved successes must be consolidated and expanded, while new comrades-in-arms and (sub)project managers must be recruited, as well as initiating further waves of transformation. Also, Key personnel must not be overburdened in the process—since, a burnout caused by the change initiative won't be helpful.

(8) Incorporating changes into the culture: It takes years for major changes to become so ingrained in the organization that they are not challenged at every opportunity. This process can be supported by targeted employee development and natural fluctuation, as well as by architectural measures and the creation of new rituals.

Although this systematic eight-step process gives the impression that large-scale change can be achieved by simply following a recipe, this is a gross misconception. However, a skilled leader can significantly boost the chances of project success by adhering to these eight points. The sequence of steps also follows a logical progression, so skipping any due to a misguided sense of urgency will likely lead to failure.

Finally, a key insight from John P. Kotter is worth quoting, especially for analytically inclined managers: "see-feel-change" is more successful than "analysis-think-change". This change principle is very much in the spirit of "go to Gemba". Don't manage change from the office and in lots of coordination meetings, but go to the people and drive change - on the ground, every day.

**Lean Implementation in an Optical Company**

Now that you have explicitly practiced all the Lean methods and are about to graduate, you are motivated to accept a position as a Lean expert in a manufacturing company. You have chosen a medium-sized company in the optical industry in Germany that, where there's a clear interest in implementing Lean. You are anticipating a challenging role with a lot of creative freedom. On your first day at work, you have an appointment with the CEO, who gives you a brief overview of the company's current status and challenges.

The traditional company, which has been in existence for over 125 years and employs 210 people, is to be transformed into a "customer-oriented and fast company". There are two production departments—one for glass lenses and another for glass prisms—and two assembly departments for saleable optical modules in the European supply business and for binoculars sold by the company itself. These productions are supported by logistics, planning, and tooling department. The company's administration handles purchasing, sales, service, and quality management. There's also a powerful development department with many technical experts. Last but not least, the company's founder, 125 years ago, was an optical specialist who held numerous patents (which of course have since expired). This also explains the traditional and technically oriented corporate culture.

The company faces increasing competition from Asia and needs to reposition itself to stay competitive. The problem is issue isn't lack of expertise but an organization that's not aligned with customer value, leading to waste across business processes. It's clear that decisive and far-reaching action is needed. The management culture needs to shift from a top-down to a team-based approach to manage this major change.

To this end, management has defined the following goals as part of their "New Beginning" strategy:

- Reduction of throughput time by a factor of 2
- Reduction of quality complaints from customers by 30%

- Reduction of production costs by 20%
- Reduction of stock and work in progress (WIP) by 70%
- Reduction of material/procurement costs by 20%.

These goals seem ambitious, and while the CEO agrees, they have to be achieved. Improving costs and increasing performance within the year is essential for the company's survival. After the meeting with the CEO, you walk into your new office with some trepidation. It feels challenging to you as a young professional with limited knowledge of the company's internal processes and technologies—and even more so to the company's proud employees. After all, with your deep understanding of Lean and change management, and you already have ideas about how you can help the company.

***Task***

*Use the following questions to structure your approach for the company in question. If you lack information, make realistic and transparent assumptions.*

1. *Roughly describe your first actions and measures in the next 4 weeks and give reasons for your approach.*
2. *What specific Lean methods are you using and why? Explain the order in which the methods will be used.*
3. *In addition to using Lean methods, what else do you as the project manager need to consider when implementing this Lean transformation and fundamental change? What are you doing regarding good project management—and why?*
4. *Develop a roadmap (a rough project phase plan) for the next 1 ½ years.*

*Management is curious about your plan of action. They expect a comprehensive presentation.*

***Solution***

1. *Actions and measures for the next 4 weeks: Familiarize yourself with the processes (possibly by assisting), assess the current situation (see methods under 2.), conduct a target gap analysis to determine how ambitious the required targets are for the company, work with the CEO and the other board members to devise a plan for creating and communicating a sense of urgency. Also, establish a change team (management coalition) and create a*

*1-min message and a roadmap ("strategy" or project phase plan) for later communication.*

2. *First, apply analytical Lean methods to address most pressing areas for improvement, including lead times, inventory on hand, and inventory levels. Value stream mapping and inventory analysis should also be employed. For machines with clear inefficiencies and/or known quality problems, determine the OEE metric and analyze the situation using a Pareto chart. The results from this waste analysis will directly flow into the development of a vision. All methods for eliminating waste become only relevant in connection with the pilot projects and with the participation of selected employees. Otherwise, A 20–30% improvement in quality and costs will likely occur as a secondary effect of the subsequent Lean measures.*
3. *Use project management techniques to start and plan the project. Clarify the time frame and available resources (including costs). The performance objectives have already been established, covering all three dimensions of the so-called "magic (or iron) triangle" of quality, time, and cost. Specifically, you need to address the following work packages: creating a stakeholder or force field analysis, developing a general communication plan, identifying and addressing project risks, laying the groundwork for project management (e.g., for progress tracking), and defining metrics to measure the success of the Lean initiatives*
4. *See the figure below for an example of a roadmap. The most appropriate employees should be fully involved in Lean activities immediately (at least during the detailed analysis). Regularly tracking and reporting project progress is also essential, especially to the workforce (not shown separately in the roadmap, as it is linked to the milestones anyway), and to management. A steering committee meeting, preferably led by the CEO or the person in charge, has proven successful. Success and progress against KPIs must be measured, but also "sold" and communicated (project marketing). Success and transparency also increase motivation.*

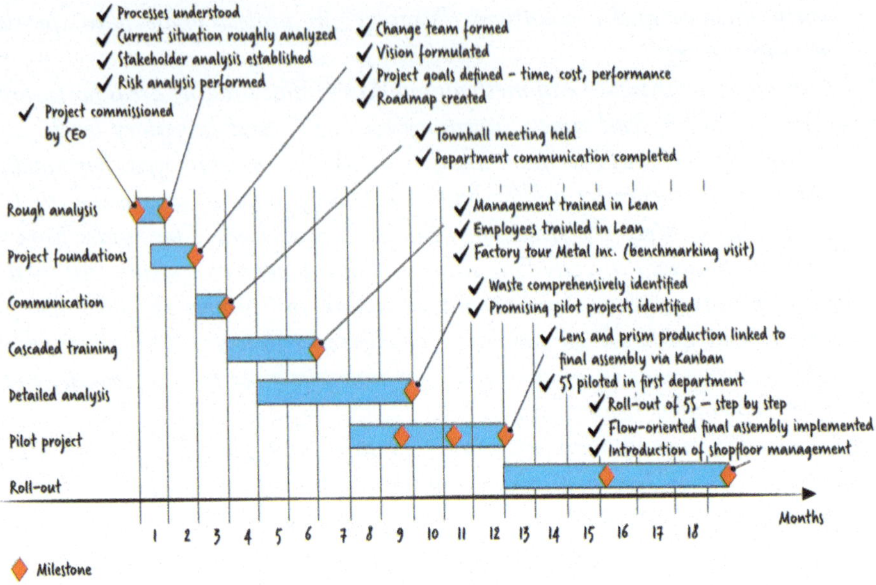


## Order Fulfillment Process at a Machine Tool Manufacturer

While Lean change initiatives are not limited to the shopfloor, they also possess significant waste in higher-level business processes and administration, where the Lean philosophy can be positively impacted. This being said, consider this exercise to deepen your understanding of change management.

Faced with an increasing challenging competition, the management of a global machine tool manufacturer based in southern Germany decided to significantly improve the order fulfillment process for standard machines at all two national and five international locations. As part of the internal benchmarking project titled "Learning from the Best", all existing processes were analyzed by a small central core team and compared with the usual key figures. The Japanese site emerged as the winner and is now to serve as a model for all other sites. The Japanese order fulfillment process is characterized by high efficiency, short lead times, and an extremely low error rate. Consequently, the management expects it to be replicated at all other sites, expecting full implementation within eight weeks in order to win back customers who have migrated, or prevent further attrition. Strictly speaking, although there appears to be a cost issue, little is known outside the inner circle of management about its relationship to the order fulfillment process.

You, as the site manager for the Heilbronn site with 750 employees and, together with the other site managers, receive the above-mentioned implementation assignment during a video conference. So far, you have had little involvement in the benchmarking project itself, contributing only flowcharts, numbers, and information about the site-specific ("home grown") IT system. However, because of your knowledge of Lean and change management, you are asked to stay on the call after the general video conference to share your ideas on how to proceed. You asked for a day to think about the situation before providing specific feedback. The following questions will help you analyze the situation:

***Task***

1. *When considering departments involved in the order fulfillment process, it's important to acknowledge that the company operates in a manufacturing network. For example, the fully assembled sensors for the machine tools produced in Heilbronn come from the plant in France, while Italy supplies all the metal housings. Also, Japan, solely manufactures all components and has no custom machines at all!*
2. *Examine the situationconsidering your knowledge of change management. What numerous problems does management face? How do you assess your situation as the site manager in Heilbronn?*
3. *Using the first four points of the eight-step approach: Assuming that the change initiative is long-term in nature, what advice do you give to management during the following videoconference?*

***Solution***

1. *Virtually all departments are involved in one way or another: sales, materials management with purchasing, logistics, production, QA, and IT. The order fulfillment process is an integrated process that runs through the entire company. The quality of this process is a reflection of the company as a whole.*
2. *There is no evidence that the "Learning from the Best" initiative has become more widely known within the company or at the sites. Employees are unaware of the company's predicament. Multi-location problem: outdated and complex production process enriched by intercultural differences. Germany is not Japan; they have different product and customer structure, different mentality, different structural requirements, and IT systems. In view of these problems, the required implementation time is more than unrealistic. The comparison helps to learn—but nothing more. The comparison with Japan does not cause concern, it only "stirs up" defensive reactions.*

3. *The following points must be considered to initiate a real change process:*
   - *There is no convincing message from management about the project's background and urgency. How big is the cost problem? There is a lack of basic figures, data, and facts that can be communicated by all site managers—butthe emotional side also needs to be considered to get people excited about the project. The entire workforce needs to understand the situation of each site and the company as a whole.*
   - *You recommend forming a change leadership team that extends beyond the immediate management team and includes all seven locations. It is important to ensure that each member of this "guiding coalition" is a good communicator, has integrity, has relevant expertise from the order fulfillment process, and is a team player.*
   - *To adequately communicate the vision of change, there needs to be a vision (operationalized in the form of a roadmap, among other things) to work toward. Communication across all locations must reach every employee.*
   - *If the change initiative is to be successful, the timeline must be fundamentally revised. First of all, enough time must be set aside to start the project properly, considering the points mentioned above. You need to create an ambitious yet realistic timeline for the change. It can be assumed that we are dealing with a Lean transformation for the future, not a survival-oriented restructing. Regarding restructuring and survival, the Lean approach is not the right approach because it is "too slow".*

# Glossary

**5M and 7M** Pre-structured Ishikawa diagram for problem cause analysis with the branches man, machine, material, method, mother earth (environment) (5 M) and management, as well as measurement (7 M).

**5R** 5-times-right basic principle or just-in-time principle means providing the right part, in the right quality, at the right time, in the right place, in the right quantity.

**5S** Lean method, which organizes the workplace in five steps: sorting, setting in order, making shine, standardizing, and sustaining the standard.

**8-step method** The 8-step method is a step-by-step procedure for implementing major change projects in change management.

**A3** Method for structuring a problem-solving process on a DIN A3 sheet of paper.

**ABC-XYZ analysis** Classification of parts according to part price for purchased parts or manufacturing costs for in-house production (ABC classification) and consumption variance (XYZ classification).

**Activity sampling study** Statistical method for determining time requirements by taking samples over a longer period of time.

**Andon** Method from the Toyota production system for immediate visualization of process problems, e.g., using a signal lamp.

**Batch size** See lot size.

**BoA** Load-oriented Shopfloor Control; determines the sequence of production orders taking into account a load limit for operations.

**BOM** Bill of Materials, parts list.

**Circular economy** The conscious handling of materials after their initial use; reuse of materials.

**CNC** Computer Numeric Control; digital control of machine tools.

**$c_p$, $c_{pk}$** The $c_p$ and $c_{pk}$ are process capability indices that show how stable and centered a process is.

© The Editor(s) (if applicable) and The Author(s), under exclusive license to Springer-Verlag GmbH, DE, part of Springer Nature 2025
R. Hänggi et al., *LEAN Production Training for Practice and Study—35 Exercises with Solutions*, https://doi.org/10.1007/978-3-662-70468-4

**C-parts** Materials with low value (such as small parts, e.g., screws, nuts).

**CIP** Continuous improvement process, also known as "Kaizen" in Japanese

**Critical path** Shortest path through the activities in a network (e.g., in lead time scheduling) without no time buffer left; if an activity on the critical path is postponed, the completion date will also be postponed.

**CT** Cycle time; time that elapses between the completion of two parts (workpieces) in a process step or workstation.

**DF** Degree of Flow; ratio between processing time and idle time of a part in the overall process.

**EPEI** Every Part Every Interval; key figure for the shortest time period until the same part is produced again.

**ERP** Enterprise Resource Planning; IT system for planning and controlling the flow of goods and values within the company.

**FIFO** First In First Out; parts produced or stored first are also removed and processed first.

**FMEA** Failure Mode and Effects Analysis; structured procedure for identifying risks and possible corrective measures.

**Gemba** Japanese for "the real place." In the Lean context, this is the place where value creation occurs.

**Handling steps** Processes such as transporting, checking, unpacking, storing or even relocating are handling steps.

**Handling step analysis** The handling step analysis shows the handling steps for a process in context and helps to identify the waste contained in the process.

**I** Inventory; triangle symbol with an inscribed "I" is often used in the value stream map.

**Idle time** Time during which a machine is not working and causes waiting for materials and workers alike.

**Industry 4.0** Use of digital technologies to increase productivity in an industrial company, especially in production.

**Ishikawa diagram** Method for the comprehensive analysis of a problem and its causes; also called fishbone diagram.

**Jidoka** Intelligent mechanical solutions to avoid errors, the path to all-encompassing automation, "Poka-Joke first, then automate."

**JIS** Just-in-Sequence, delivery of parts or assemblies in a specified sequence.

**JIT** Just-in-time, direct delivery between processes without intermediate storage/retrieval process.

**Johnson algorithm** The Johnson algorithm is an optimization method for sequence planning.

**K** Number of Kanban cards or containers.

**Kaizen** Management approach for continuous improvement.

**Kanban** Methods for linking parts production/supply with the parts consumption of downstream processes by means of a signal (Kanban); thus, implementing pull control.

**KPI** Key Performance Indicator; measurement parameter for key metrics in the company.

**Lead time** In a generic sense, the term is used for the time that elapses between a customer placing an order and the arrival of the shipping. Whenever used internally, it is the time it takes for a part to pass through a process or a series of related processes; see "throughput time."

**Lead time scheduling** A planning method for defining start and end dates for production processes in in-house production.

**Lges** Total travel distance, e.g., when evaluating a spaghetti diagram.

**LOOR** Load-oriented Order Release determines the release of production orders while consideringt load limits of operations (German term: BOA—Belastungsorientierte Auftragsfreigabe).

**Lot size** Number of identical parts produced or procured together.

**Production Scheduling** Planning department for production orders for in-house production; often also referred to as production control.

**LLC** Large load carrier.

**Make-to-order** Control type of a production order in which products are manufactured to (customer) order and not to stock.

**Make-to-stock** Control type of a production order in which products are manufactured to stock.

**Material flow matrix** Shows the relationship between sources and sinks of the material flow in production.

**Milkrun** Clocked tugger (or logistics) trains to supply consumption locations in production with new parts at fixed intervals.

**MRP** Material Requirements Planning; term for material requirements planning via BOM explosion and stock reconciliation.

**MRP II** MRP including capacity planning.

**MTM** Methods Time Measurement; method for determining process times using time modules with standardized and defined specifications.

**Muda** Japanese term for waste.

**OBC** Operator Balance Chart; diagram for visualizing the distribution of work, in Japan this method is called Yamazumi Board.

**OEE** Overall Equipment Effectiveness; method for calculating system or process effectiveness.

**One-piece flow** One-piece production; also referred to as "lot size 1 production."

**PT** Processing time, or time per unit to process a product.

**Pareto chart** Visualization of the 80/20 rule in the form of a bar chart sorted according to the frequency of occurrence of e.g., errors.

**Pareto rule** 80/20 rule means that 20% of the causes/issues cause 80% of the problems/results.

**PDCA** Plan-Do-Check-Act; approach from quality management for continuous problem solving in four phases.

**Poka-yoke** Any device for avoiding errors and human mistakes by technical or organizational measures, either in the process or on the product.

**PPM** Defective Parts per Million Parts; can also be abbreviated to DPPM.

**PPS system** Production planning and control system; computer program or system that plans and controls production centrally according to the push logic.

**Predictive maintenance** Maintenance ahead of time, especially of production machines, to avoid unexpected machine breakdown; relevant use case for learning from data in the factory of the future.

**Pull** Parts replenishment is always triggered by customer consumption.

**Push** Centrally planned production orders, independent of the maximum stock level and the actual parts consumption of downstream processes.

**$Q_{opt}$** Optimum lot size based on calculations like the Andler formula.

**Qty** Quantity.

**Rank order clustering algorithm** Algorithm for the optimal allocation of parts to a production cell with the goal of minimizing routes.

**REFA** Time calculation method of the Association for Work Design, Organization, and Corporate Development.

**RPN** Risk priority number as part of an FMEA; product of the probability of occurrence of a failure, its probability of detection, and the corresponding degree of severity.

**Setup time** Time to prepare a machine or process for the actual production/assembly process.

**Sequencing** Sequencing means determining the succession of work orders.

**SFM** Shopfloor management; management at the location of value creation. The status of production is discussed in short, standardized daily coordination meetings and—if necessary—measures are defined to meet targets.

**Safety stock** The safety stock is the "emergency stock"—it is intended to cover unforeseen events and is therefore not included in the MRP run.

**SLC** Small load carrier.

**SMED** Single Minute Exchange of Die; method for setup time reduction with the goal of carrying out a changeover process "in the single-digit minute range."

**Spaghetti diagram** Graphical representation of routes at workplaces; the representation of complex routes resembles "spaghetti."

**Swim lane diagram** Representation of a flowchart with (swim) lanes running from left to right, which stand for roles in the process or departments, and in which the activities or processes carried out are positioned.

**Throughput time** The time it takes for a part to pass through a process or a series of related processes. This is another word for cycle time.

**TT** Customer Takt Time; cycle time derived from customer demand.

**UTL and LTL** Upper Tolerance Limit and Lower Tolerance Limit of a relevant process parameter.

**Waste** Excessive consumption or inefficient use of resources in processes (Muda).

**Waste Walk** Structured inspection of production to identify waste.

**WT** Available Working Time of processes

**Waiting time** Time a part spends in the material buffer in front of a workstation; sometimes "idle time" is used as a synonym.

**Yield** Ratio of good parts to all parts produced as a percentage, or final yield.

**Zoning** Arrangement of materials, work equipment, or storage locations according to access frequency and priority.

# References


Budde, L, Hänggi, R., Friedli, T., Rüedy, A., (2023). *Smart Factory Navigator: Identifying and Implementing the Most Beneficial Use Cases for Your Company — 44 Use Cases That Will Drive Your Operational Performance and Digital Service Business.* Springer Cham. https://doi.org/10.1007/978-3-031-17254-0

Dennis, P. (2016). *Lean production simplified: A plain-language guide to the world's most powerful production system* (3rd ed.). CRC Press, Taylor & Francis Group.

Hänggi, R., Fimpel, A., & Siegenthaler, R. (2021). *LEAN Production – einfach und umfassend: Ein praxisorientierter Leitfaden zu schlanken Prozessen mit Bildern erklärt*. Springer Berlin Heidelberg. https://doi.org/10.1007/978-3-662-62702-0

Imai, M. (1986). Kaizen (Ky'zen), the key to Japan's competitive success. New York: Random House Business Division. http://archive.org/details/kaizen00masa

King, J. R. (1980). *Machine-component grouping in production flow analysis: An approach using a rank order clustering algorithm.* International Journal of Production Research, 18(2), 213–232. https://doi.org/10.1080/00207548008919662

Klevers, T. (2007). *Wertstrom-Mapping und Wertstrom-Design: Verschwendung erkennen, Wertschöpfung steigern.* MI.

Klevers, T. (2015). *Agile Prozesse mit Wertstrom-Management: Ein Handbuch für Praktiker; Bestände abbauen, Durchlaufzeiten senken, flexibler reagieren* (2., überarb. Auflage). CETPM Publ.

Koch, A. (2021). *OEE für das Produktionsteam: Das vollständige OEE- Benutzerhandbuch – oder wie Sie die verborgene Maschine entdecken* (C. May, Ed.; 4., korrigierte Auflage). CETPM Publ.

Kotter, J. P. (1996). *Leading change.* Harvard Business School Press.

Linß, G. (2018). Qualitätsmanagement für Ingenieure (4. akt. u. erw. Aufl.). Carl Hanser, ISBN 978-3-446-43936-8.

Modig, N., & Åhlström, P. (2017). *This is lean: Resolving the efficiency paradox* (1st ed.). Rheologica Publishing.



© The Editor(s) (if applicable) and The Author(s), under exclusive license to Springer-Verlag GmbH, DE, part of Springer Nature 2025



R. Hänggi et al., *LEAN Production Training for Practice and Study—35 Exercises with Solutions*, https://doi.org/10.1007/978-3-662-70468-4

Ohno, T. (1988). *Toyota Production System: Beyond Large-Scale Production* (1st ed.). Productivity Press.

Peters, R. (2017). *Shopfloor Management: Führen am Ort der Wertschöpfung*. LOG_X Productivity Press Development Team. (2020). *Kanban for the Shopfloor*. Productivity Press. https://doi.org/10.4324/9780429273063

Roser, C., & Shook, J. (2022). *Alles über Verbrauchssteuerung: Planung, Umsetzung und Pflege von Kanban, CONWIP und anderen Pull-Systemen in der Schlanken Produktion* (Version 1.00, 10.01.2022, Erste deutsche Ausgabe). AllAbout-Lean.com Publishing.

Rother, M., & Shook, J. (1999). *Learning to see: Value-stream mapping to create value and eliminate MUDA*. Lean Enterprise Institute.

Schönsleben, P. (2020). *Integrales Logistikmanagement: Operations und Supply Chain Management innerhalb des Unternehmens und unternehmensübergreifend*. Springer Berlin Heidelberg. https://doi.org/10.1007/978-3-662-60673-5

Seghezzi, H. D., Fahrni, F., & Friedli, T. (2013). *Integriertes Qualitätsmanagement: Das St. Galler Konzept* (4., vollst. überarb. Aufl.). Hanser.

Shingō, S. (1996). *Quick changeover for operators: The SMED system* (Productivity Press, Ed.). The Press.

Sobek, D. K., & Smalley, A. (2008). *Understanding A3 thinking: A critical component of Toyota's PDCA management system*. CRC Press.

Suzaki, K. (2014). *New shopfloor management: Empowering people for continuous improvement*. Free Press.

Teeuwen, B., & Grombach, A. (2019). *SMED: Die Erfolgsmethode für schnelles Rüsten und Umstellen* (3., unveränderte Auflage). Deutscher Management Verlag.

VDI-Gesellschaft Produktion und Logistik (2022). VDI-Richtlinie 5586 Blatt 2, *Routenzugsysteme – Planung und Dimensionierung*. Beuth Verlag, Berlin.

Wiendahl, H.-P., & Wiendahl, H.-H. (2019). *Betriebsorganisation für Ingenieure* (9., vollst. überarb. Aufl.), Hanser.

Wildemann, H. (1998). *Die modulare Fabrik: Kundennahe Produktion durch Fertigungssegmentierung* (5., überarb. und erg. Aufl). TCW-Transfer-Centrum.

Womack, J. P., Jones, D. T., & Roos, D. (1990). *The machine that changed the world: Based on the Massachusetts Institute of Technology 5-million dollar 5-year study on the future of the automobile*. Rawson Associates.

The manufacturer's authorised representative in the EU is Springer Nature Customer Service Centre GmbH, Europaplatz 3, 69115 Heidelberg, Germany. If you have any concerns regarding our products, please contact ProductSafety@springernature.com

Printed and bound by CPI Group (UK) Ltd, Croydon, CR0 4YY
30/09/2026
02225474-0007